S0-ATS-017

Advance praise for *The Genome Generation*

Elizabeth Finkel is the rare author who conveys complex science in understandable and thrilling ways, but is never condescending. *The Genome Generation* helps us understand who we are, how we got here and why we do what we do. Most impressively, as we move into the heart of the genomic revolution, Finkel provides a guide as to where we are going. This is an accomplished work of scientific literacy.

Jon Entine, Director, Genetic Literacy Project, George Mason University

Finkel humanises an otherwise scientific and technical tale, tracing in simple and engaging language the human quest to understand genetics and how this has impacted on everyday lives in areas from food security and agricultural production to public health and evolution. She describes the age of the genome in a way that is both current and thoroughly enjoyable.

Thomas A. Lumpkin, Director-General, International Maize and Wheat Improvement Center

Ella Finkel presents a fast-paced, anecdotal history of the workings of the human genome. She races past genes, their isolation and control, the way they evolve, and some hints on how they mediate health and disease. She is at home with the genetics of AIDS, the coat colour of a mouse, or the genes behind a food revolution. Her book is a great read, always positive, with lots of references for those who want to dig deeper, and full of personal reminders that scientists are real people who often live for the excitement of a research finding.

Robert Williamson, Policy Secretary, Australian Academy of Science

WITHDRAWN

The Genome Generation is a delightful and engrossing synthesis and weaving of concepts from wide-ranging and disparate sources, and really captures the quirky nature of science and the joy of discovery.

Richard Roush, Dean, Melbourne University School of Land and Environment

For too long 98% of our genome was dismissed as useless 'junk'. In *The Genome Generation* Elizabeth Finkel meets the mavericks who were right—scientists prepared to delve deeper into one of the greatest scientific orthodoxies of our time. These tales from the genetic frontier leave you with no doubt life as we understand it is about to look very different, with dramatic implications for what we eat, how we heal and who we are.

Natasha Mitchell, Presenter, ABC Radio National

The Genome Generation is absolutely riveting. It captures the promise of the exciting new universe of coded information arising from the reading of the genome in a style that is magical and an insight that is deep. The author's ability to convert extreme complexity into utter simplicity is amazing. It was a lifetime experience for an individual like me with only a nodding acquaintance of modern biology to grasp the essence of the concepts ranging from 'central dogma' to 'junk gene' so painlessly. These tales from the frontier are a 'must read' for everyone who wishes to understand our past—the logic of evolution—or take a peep into our exciting future at the creation of 'superplants' through 'digital agriculture'.

R.A. Mashelkar, CSIR Bhatnagar Fellow and President,
Global Research Alliance

Dr Finkel definitely has a way with words, and she has again put together a most wonderful tale of one of the most incredible and important achievements of biological and medical research: deciphering and applying the discoveries of genome science. A wonderful introductory chapter gets the reader ready for a fantastic journey on the nature and history of the gene, with all of the right players and discoveries presented. From figuring out the genetic basis of human disease to proposing new ways to feed all the people on our planet, Dr Finkel clearly explains genetics and relies on all of us as the genome generation to help explain hereditability and how studying plant and animal genomes will unquestionably continue to improve the human condition.

Dennis A. Steindler, Professor of Medicine, The Evelyn F. and
William L. McKnight Brain Institute

WEST VALLEY COLLEGE LIBRARY

THE
GENOME
GENERATION

ELIZABETH FINKEL

MELBOURNE
UNIVERSITY
PRESS

MELBOURNE UNIVERSITY PRESS
An imprint of Melbourne University Publishing Limited
187 Grattan Street, Carlton, Victoria 3053, Australia
mup-info@unimelb.edu.au
www.mup.com.au

First published 2012
Text © Elizabeth Finkel, 2012
Illustrations © Kate Patterson, 2012
Design and typography © Melbourne University Publishing Limited, 2012

This book is copyright. Apart from any use permitted under the *Copyright Act 1968* and subsequent amendments, no part may be reproduced, stored in a retrieval system or transmitted by any means or process whatsoever without the prior written permission of the publishers.

Every attempt has been made to locate the copyright holders for material quoted in this book. Any person or organisation that may have been overlooked or misattributed may contact the publisher.

Typeset by TypeSkill
Illustrations by Kate Patterson, MediPics and Prose
Printed by Griffin Press, South Australia

National Library of Australia Cataloguing-in-Publication entry

Finkel, Elizabeth.
The genome generation / Dr Elizabeth Finkel.

1st ed.

9780522856477 (pbk.)
9780522860313 (ebook)

Includes index.

Human gene mapping.
Human genetics.

599.935

FSC
www.fsc.org
MIX
Paper from
responsible sources
FSC® C009448

QH
447
·F56
2012

I dedicate this book to the memory of my beloved parents,
Leon and Dora Szer.

And to Kagiso, a boy brimming with potential and a thirst
for knowledge, whose life was cut short by AIDS.

2/20/14   32.40

Contents

Acknowledgements

They say it takes a village to raise a child. Writing a book is very much like raising a child.

If I am the conceiving mother who gestated it these three-and-a-half years, then its father is my husband Alan Finkel, who applied discipline and rigour, and nurture. Rachel Nowak, former Australasian editor of *New Scientist*, was the skilled midwife. When the baby was stuck in the birth canal, her brilliant editing smoothed the passage. And it has had two wise elder brothers—my sons Victor and Alex, who indulged, played with, and disciplined the child as necessary. A nurturing grandmother, Vera Finkel, gave encouragement at all the right times.

Then there have been my readers, mostly dedicated family and friends who have listened tirelessly to my ravings and read my drafts.

Thank you to Tamara Bruce, Eva Bugalski, Kerry Bugalski, Christine Copolov, David Copolov, Ros Gleadow, Debbie Grace, Kirsty Hamilton, Mira Hetnal, Brinley Hosking, Lea Jellinek, Geoffrey Kempler, Harry Kestin, Ron Lazarovits, Robert Lefkovic, Emily Purcell, Patricia Rich, Ruth Rosen, Nadia Rosenthal, Tom Schwarz and Stephanie Wayne.

And there must be close to a hundred scientists whom I have pestered these past three-and-a-half years. I thank them all, and especially a few who became mentors. My heartfelt thanks to: Eldon Ball, T.J. Higgins, Sharon Lewin, John Mattick, David Miller, Vicki Pierce, Richard Richards, Roger Short, Catherine Suter, Peter Visscher and Robert Williamson.

In the last phase of rendering this book, I needed the help of artists.

Christine Zavod captured my feeling for wheat and rice plants with her lovely paintings.

Kate Patterson took my material one step further into the graphic realm with her lucid cartoon sketches.

Finally, the staff of MUP have been a pleasure to work with. Thank you to Foong Ling Kong for supporting the first three years, to Kate Indigo for unwavering understanding and support through the taxing final editing phase, and to Cathy Smith for bringing everything to fruition with such good humour and aplomb.

Introduction

What's a genome, you say?

Imagine you are an android. I come along with a tiny gold screw-driver, unscrew your nipple screws and out slides a tray with your gleaming hard disc on it. I pop that disc into a computer and start reading the code that makes you you.

That's your genome. There are three billion DNA letters in the code, and there are two copies of it, one from Mum and one from Dad, which is a lot of code to read. The first full reading of a human genome was accomplished in February 2001. It was a monumental effort, taking eleven years, three billion dollars and an army of researchers, mostly in the US and UK.[1]

When it was finally achieved, the technological tour de force was hailed as the 21st century's equivalent of putting a man on the moon.

So, just over a decade on, you might ask: Well, so what? What has the impact of the human genome project been?

That's the question I set out to answer with this book.

The human genome project hasn't put us on the moon; it has launched us into a new universe. Some critics gripe that the genome didn't deliver. Indeed we haven't yet cracked the mystery of how genes create common diseases or make us intelligent, beautiful or bad. Some say we may never get to that point, because what we have is a complex system.

The reading of the genome has given us a new universe of coded information to explore. The new generation of geneticists don't wield

test tubes; they are computer geeks. And they haven't just been reading human genomes; they've been reading the genomes of everything
from sponges to chimpanzees.

This new universe is full of weird and wonderful stuff. Many of
our scientific dogmas have toppled. We can't easily define a gene any
more; we admit most of our genome is doing stuff akin to a high-level
software program that we have yet to decode, and we are at a loss to
explain why we have the same number of genes as a roundworm. And
horror of horrors, a laughable theory has raised its head out of the
mire: Lamarckism, the idea that experience can influence your genes
and those of your offspring, is back.

After decades of dogma there is a new and dizzying wind of openness
among researchers. 'I promised myself that from now on any bizarre
finding in my lab will always be treated with respect, even though it
does not make much sense,' Jean-Michel Claverie, a professor of medical genomics and bioinformatics at the University of Mediterranée
School of Medicine in Marseilles told me. Mark Mehler, a professor
of neurology and psychiatry at Albert Einstein College of Medicine in
New York, wonders if scientific paradigms will ever be able to solidify
again. 'Given that the change we are going through is so cataclysmic,
is beyond what we've ever seen, can any scientific paradigm sustain
this scope of shift? Will we keep a unity of thought?' He is collaborating with the University of Queensland's John Mattick, the doyen of
the genome's software, to discover how the genome creates cognition.[2]

Resourceful scientists are harnessing the power of what they are
finding. Learning how the software of the genome operates promises
to give us a nifty new class of drugs known as RNAi. In agriculture,
getting under the bonnet of plants is letting us tinker with their parts
as never before. It took us ten thousand years to breed a scrawny
Middle Eastern weed into wheat. But we only have 40 years now to
feed a population heading for nine billion. Scientists are speeding up
traditional breeding to create super plants.

Researchers are also mining the human genome to find varieties of genes that either predispose us to disease or protect us. For instance, HIV researchers have mined the genomes of people who successfully fight the virus, to discover new strategies to fight HIV. While in common diseases such as diabetes, heart disease or schizophrenia, researchers have gone mining to see if they can discover the predisposing genes. So far the findings have been limited. But all that is set to change as the mining operations get into high gear. Until now we only sampled human genomes at conspicuous landmarks—sites where the code is likely to vary from person to person, called common SNPs. Now the cost of reading every single letter of the genome has crashed. What once cost $3 billion now costs less than $10,000. Soon it will be $100. In three to five years, researchers will have read genomes from tip to tail. Supercomputers and computer nerds will be revealing all their secrets.

It doesn't necessarily we mean we will get to the point where knowing your genome is knowing your destiny. This is a complex system, both complex within itself and complex in its interactions with the environment and experience. But it probably means we are getting a lot closer to having to reckon with genetic destiny.

Reading the genomes of creatures up and down the evolutionary tree has launched the next chapter of our understanding of evolution. Darwin gave us natural selection, Mendel gave us genes; and the two were married in the 1930s with the modern synthesis. Now we read genomes, the ultimate way to probe the ancestry of life on earth. And just as the manual of a machine reveals the logic of its design, so the genetic manual is revealing the logic of evolution. It has ushered in what evolutionary biologist Eugene Koonin at the National Center for Biotechnology Information in Washington, has dubbed the 'post-modern synthesis'. There have been many shocks: it's hard for scientists to explain why a sponge turns out to have most of the genes it needs to make a brain.

The reading of genomes has revolutionised every aspect of the biological sciences. The stories told here are exemplars. I travelled through space and time to find them—to the beginnings of our 'idea of a gene' to meet the great pioneers (and was surprised to find many of them were physicists on a quest to discover the simple law that would explain the mystery of life), to Botswana and Boston to learn about HIV, to Mexico to meet Norman Borlaug—the father of the Green Revolution—to wheat fields in Warracknabeal and Leeton in rural Australia, and to Townsville at the edge of the Great Barrier Reef to visit a lab that is reading the genome of coral.

Here are the stories that have been distilled from those travels. They are my answer to the question: What does it mean to be a part of the genome generation?

1

The Idea of a Gene

I'm sitting around a table with a group of academics from a university arts department. It's a formal meeting. Each academic takes a turn to speak. Then it's the turn of a thick-set dark-haired fellow. He's exceptionally articulate and I enjoy the virtuosity of his language. I really can't remember what he said, but I do remember a phrase that stunned me. Leaping out from amid the artsy words came: 'It's in the institutional DNA.'

Wow, I thought, DNA has become a cliché.

The virtuoso speaker was using DNA to connote something innate and defining. DNA is the physical substance that makes up a gene, a word that is also a cliché. My speaker might equally have said, 'it's in the institutional genes'. Indeed, you hear it all the time now: 'It's in his DNA,' someone quips about President Barack Obama's stance on civil rights. Do these speakers realise that in their short clichés they are telescoping a 2,300-year intellectual quest?

The idea of a gene—the inborn factor that makes Obama Obama—has tremendous resonance. And not just for this generation. Aristotle mused about how the acorn was 'informed' by the plan of an oak tree or how the egg carried the 'concept' of a chicken.[1] From Aristotle's musings to the discovery of the DNA double helix in 1953, scientists have been on a quest to discover the physical identity of a gene.

The quest shows no signs of having ended. Since 2001, we have been able to read the human genome in all its glory: three billion letters of DNA that make us who we are. As 21st century geneticists grapple with the revelations, our idea of a gene is once again undergoing a major overhaul.

**

Aristotle had an idea of a gene. But it was the monk Gregor Mendel who gave the idea substance in the 1860s. Whilst performing his monkish duties in what is now the Czech Republic, Mendel bred peas and like many breeders before him, was struck by the heritability of traits.

Breeders had taken advantage of heritability for millennia—that's how we got Chihuahuas from wolves.[2] The remarkable thing about the scientifically minded Mendel was that he uncovered the rules of the breeding game. For instance, when he crossed a yellow pea to a light green one, all the offspring came out yellow. Mendel realised that yellow skin was 'dominant'. But though yellow skin might dominate light green skin, it didn't obliterate it. When Mendel crossed the yellow offspring back to each other, he got back some green skins: about a quarter of the offspring were light green.[3] It happened every time—a quarter of the offspring were throwbacks. The light green peas—the 'recessive' trait—came back as pristine as blue-eyed babies springing from the loins of brown-eyed parents. Mendel realised that hereditary factors, or genes as they were later to be called, didn't mix or dissolve; they were as insoluble as glass beads. Moreover, they came back in a predictable mathematical ratio.

Mendel did not coin the term 'gene' to describe his heritable factors. In fact, Mendel went largely unrecognised in his lifetime, most notably by Charles Darwin, who could have greatly profited from his insights. In 1900, sixteen years after his death, three scientists rediscovered Mendel's laws. One of them, Dutch botanist Hugo de Vries, had previously coined the term 'pangenes' as the units of heritable material in a book published in 1889. However, the final naming rights go to Danish botanist Wilhelm Johannsen, who coined the term 'gene' in a book he published in 1909.[4] But he meant the term to be taken rather loosely. 'The word "gene" is completely free of any hypothesis,' he cautioned.[5]

Mendel the breeder took the first step towards solidifying the concept of a gene. Scientists peering down microscopes took the next. In 1902, Walter Sutton in America and Theodor Boveri in Germany observed plant cells dividing. Just before they separated, dark threads called chromosomes appeared that then split themselves between each daughter cell. Boveri and Sutton guessed they were staring at Mendel's heritable factors.

Over the next couple of decades, American scientists firmed up the evidence that chromosomes were indeed the repository for genes. Thomas Hunt Morgan and colleagues at Columbia University in New York showed that when bits of a fruit fly's chromosomes were damaged, the flies inherited altered traits—white eyes rather than red, for instance.[6] Barbara McClintock at Cornell University showed the same thing in maize chromosomes.[7]

A quirk of nature led some researchers to believe they could even *see* the genes. Before fruit fly grubs go into their cocoon stage, they need to stock up. The chromosomes of their salivary glands oblige by copying themselves a thousandfold and turning into extremely fat chromosomes. Viewed under the microscope they display an intriguing banding pattern. Researchers couldn't help but wonder: Were these bands genes? (Decades later researchers showed that they corresponded to clumps of genes.)[8] All this helped solidify the view that genes were a discrete physical entity that lay along the length of the chromosome like glass beads on a string.

With chromosomes established as the repository of genes, it was time to drill down to see what genes were made of and how they worked. The mystery that had captivated Aristotle was an irresistible temptation to physicists and from the 1930s they came galloping into biology to solve it. Part of the reason may have been their pervasive belief that 'there was nothing left to figure out in physics'.[9] From the cosmic to the subatomic, by the 1930s they had distilled the complexity of the universe into a set of elegant laws. Newton's laws described moving bodies on earth and Maxwell's laws united the forces of electricity and magnetism. Einstein's theory of relativity explained the motions of galaxies and then with $E = mc^2$ he revealed that matter and energy were two sides of the same coin. Finally the likes of Erwin Schrödinger, Werner Heisenberg and Paul Dirac cracked the quantum laws that ruled the subatomic realm.

The physicists who came to solve the mystery of the gene were hungry to discover new laws; they wanted to crack the mystery of the 'atom of the cell'.[10] Among them was Max Delbruck, a German quantum physicist.[11] In 1937 he re-invented himself from a physicist in fascist Germany to a biologist at the California Institute of Technology.[12] His physicist's training had taught him to investigate a complex problem by reducing it to its simplest form. So he chose to investigate the simplest life form known: a virus that preys on bacteria known as the bacteriophage or literally 'bacteria eater'. The bacteriophage was perched on the very cusp of what could be considered alive. Under an electron microscope, it looks like a lifeless miniature space ship; it doesn't move, perform chemical reactions or reproduce. But pour some bacteriophage on a colony of bacteria and they spring into action. Like tiny hypodermic syringes they inject their contents inside and within the space of 20 minutes, hundreds of new bacteriophage burst out of their hapless host. How a clump of molecules (mostly protein but also DNA) could achieve this feat was a complete mystery. As Delbruck put it, 'Certain large … molecules … possess the property of multiplying within living organisms, [a process] at once so foreign to chemistry and so fundamental to biology.'[13]

Delbruck was convinced that solving this mystery of multiplying molecules would uncover a new fundamental law of physics—the secret of life no less. He was right. The bacteriophage did ultimately help nail the physical identity of the gene. It also breathed life into a new field of study—the study of living molecules or molecular biology.[14] But he and other physicists also got some things spectacularly wrong. For instance, in the guessing game over the chemical identity of the gene, DNA and protein were even contenders. Delbruck and Schrödinger, then based at the Dublin Institute of Advanced Studies, bet on proteins. Delbruck even called DNA a 'stupid molecule' because it was just a very long polymer composed of four monotonously repeating units called bases.[15] Proteins, on the other hand, had complex three-dimensional structures and were made up of combinations of twenty different amino acids. As candidates for the molecules of life they were definitely the front-runners.

Schrödinger offered an elegant and inspirational meditation on the subject of the gene in a series of 1943 lectures titled 'What is Life'.[16] He mused that quantum theory, which showed that subatomic particles could exist only in discrete energy states, was also the key to the mystery of the particulate gene. A gene, he theorised, was 'an aperiodic crystal', most probably a protein, which could shift between a limited number of quantum states. And he speculated that 'Mutations are actually due to quantum jumps in the gene molecule',[17] and marvelled that the revelation of the underpinning principles of both matter and genetics had dovetailed in the same year. Max Planck's quantum physics had been published in 1900, the same year Mendel's laws of the particulate gene were rediscovered. 'Thus,' wrote Schrödinger, 'the births of the two great theories nearly coincide, and it is small wonder that both of them had to reach a certain maturity before the connection could emerge.'[18] Schrödinger's words were so elegant and inspirational, so imbued with the sense that the revelation of the great mystery of life was around the corner, that they launched many a young scientist onto the quest. Among them James Watson and Francis Crick.[19]

Notwithstanding all this, Schrödinger himself was on the wrong track. The gene was not made of protein.

**

'Stupid' DNA proved to be the physical substance of the gene. Oswald Avery and colleagues at the Rockefeller Institute Hospital in New York in 1944 proved it when they changed a harmless bacterial strain into one that caused deadly pneumonia by dousing it with the DNA of the deadly variety. However, despite this experiment many remained sceptical that the gene could be composed of DNA.[20] It took eight years and a second experiment to convince them. In 1952, Alfred Hershey and Martha Chase at the Cold Spring Harbor Laboratory in New York showed that bacteriophage actually left all their protein bits behind when they injected themselves into a bacterium. DNA alone was sufficient to instruct the copying of phage particles. DNA was the molecule of life.

DNA also fired the imagination of the young James Watson. As a PhD student in 1948, he got in with the 'bacteriophage group' in Salvador Luria's lab at Indiana University. Luria and Delbruck, both refugees from fascist Europe (Luria from Italy), were the gurus of bacteriophage research.[21] Watson was also familiar with Avery's experiment and convinced that the gene was made of DNA. And he was intrigued. DNA was composed of only four different molecules, probably arranged in a repeating string. How could these four molecules come together to create a gene, an entity that could carry the information for everything from eye colour to intelligence, and that could also copy itself?

A new technique known as X-ray crystallography offered the possibility of 'seeing' how the four molecules were arranged in three-dimensional space. The principle was something like shadow puppetry, where you contort your hands in front of a light beam to get, for instance, a rabbit shadow. In the case of X-ray crystallography, you put a molecule in the path of an X-ray beam. As long as the molecule had an orderly or crystalline structure it would generate a pattern. The trick was to work backwards from the shadow pattern to figure out the structure of atoms in your molecule. It had been used to great effect to work out the arrangements of atoms in salt crystals composed of a few atoms.[22] But at that time researchers were just beginning to try the technique with complex protein crystals such as haemoglobin (the protein that ferries oxygen through blood), which was composed of 10,000 atoms! There had been better success with fibrous proteins that naturally formed repeating structures, such as the keratin in a strand of hair. Indeed a single hair, clamped taut, could produce nice images in the X-ray machine.[23]

One of the best places in the world to learn how to do X-ray crystallography was the Cavendish Laboratory in Cambridge. The director, Sir Lawrence Bragg, had won a Nobel Prize 40 years before for pioneering the method to work out the structure of salt crystals. Now his laboratory was trying its hand at proteins.[24] And there was an added bonus. At nearby King's College in London, Maurice Wilkins was producing X-ray images of DNA, taking advantage of the fact that DNA extracted from a calf thymus formed long threads. 'I had spun a very thin fibre of DNA, almost invisible, like a filament of spider web,' recalled Wilkins.[25]

Like the keratin protein in hair, Wilkins found that these moistened threads of DNA, when clamped taut like violin strings in a bow,[26] produced striking patterns in the X-ray beam. It was these images, seen at a conference in Naples, that had turned Watson on to X-ray crystallography. As Watson wrote in his memoir: 'Before Maurice's [Wilkins's] talk I had worried about the possibility that the gene might be fantastically irregular. Now however, I knew that genes could crystallize; hence they must have a regular structure that could be solved in a straightforward fashion.'[27]

Watson would have liked to have enlisted with Wilkins but with no credentials as a chemist, lost his nerve.[28] And so in 1951, after a good word put in for him by his former boss, the 23-year-old Watson arrived at the Cavendish lab, eager to learn the art of X-ray crystallography so that he might somehow apply it to DNA.

The Cavendish, however, was not particularly interested in DNA; they were fixated on the proteins haemoglobin and myoglobin (which stores oxygen in muscle). But Watson was able to infect at least one other person with DNA fever. And that was the highly infectable, intellectually rapacious physicist Francis Crick, 12 years his senior. Crick's colleagues feared his insatiable appetite. He had a way of snatching the solution to other people's scientific problems, rather like 'doing someone else's crosswords', observed Bragg.[29] But going after the structure of DNA was a meaty enough pursuit to satiate Crick. He and Watson were soon spending every spare minute on the problem using whatever clues they could get their hands on.

The best clues were from Maurice Wilkins and his uncooperative colleague, Rosalind Franklin, about whom much has been written.[30] In Wilkins's view, Franklin had been hired to assist him; in Franklin's view the DNA project belonged to her. Watson and Crick managed to persuade Wilkins to show them the latest of the DNA images being produced and ultimately Wilkins accepted the pair as collaborators. Not so for Franklin; she was possessive about her hard-won images and preferred to ponder their meaning alone and in her own good time—to her disadvantage. Cracking the X-ray structure of DNA was a tough call. 'She needed a collaborator,' her colleague Aaron Klug

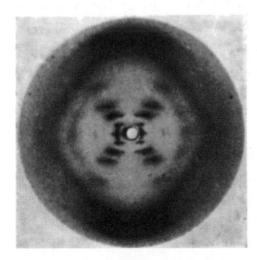

Photo 51

This X-ray image produced by Rosalind Franklin helped nail the structure of DNA (Courtesy Nature Publishing Group)

observed.[31] It was to take method, madness and a lot of collaboration to crack the structure of DNA.[32]

To get an inkling of the difficulty, imagine a rabbit shadow and how hard it would be to work out the exact configuration of the contorted hands and fingers from the image on the wall. For images made by protein crystals or DNA fibres, there were two ways to approach the problem. You took a lot of photographs at different angles and under different conditions and used mathematics to navigate back from the pattern to the arrangement of the atoms that had created it. Even with Bragg's law to guide you, it was a diabolical task. It took Max Perutz and John Kendrew over 40 years to crack the structure of haemoglobin in this way.[33]

Or you could do some inspired guesswork. The wizardly Linus Pauling, one of the greatest chemists of all time, who used quantum physics to explain chemical bonds, was the doyen of this approach.[34] When it came to divining the structure of proteins, Pauling divided them up into smaller modules called peptides. Knowing the structures of peptides, he played with ball-and-stick models of them as if they were tinker toys. One of the arrangements he came up with was the so-called alpha helix. It turned out to be spot on. Most proteins do indeed carry long stretches of alpha helix as part of their structure.

So Watson and Crick went for the inspired guesswork approach. As Watson put it, they would 'imitate Linus Pauling and beat him at his own game'.[35] It involved endless tinkering with the four building blocks of DNA—the bases adenine, thymine, cytosine and guanine (abbreviated as A, T, C and G). They were called bases because they had the chemical properties of a base rather than an acid. Their structures were known, so it was a matter of tinkering with them, trying various configurations to see one that produced the best fit.

They tried stringing the bases out in a helical configuration like beads on a twisted string. It had worked for proteins, why not for DNA? However, helices come in different varieties. For one thing, they can be single as in the protein alpha helix but they could also be double, like a spiral staircase, or even triple. (Pauling was to wrongly propose that DNA formed a triple helix.)[36]

A clue was provided by another escapee from Nazism, Erwin Chargaff, a Jewish Austrian biochemist who found refuge at Columbia University in New York. Chargaff had chopped DNA up into its four components and reported an intriguing finding. The *quantities* of the bases were curiously matched: the amount of adenine always equalled the amount of the thymine, and the amount of guanine always equalled the amount of the cytosine. It didn't matter whether the DNA came from a wheat seed or herring sperm, the pairing held up.

Watson and Crick had been toying with a double helix: two strings of spiralling bases lying alongside each other. Perhaps the bases from each helix reached out to join each other like the rungs of a spiral ladder. They supposed that like paired with like, A reaching out to A, G to G, C to C, T to T.

They were very close but they were wrong.[37] In a moment of epiphany, Chargaff's data showed them what the true pairing had to be. If the amount of A always equalled the amount of T, and the amount of G always equalled the amount of C, then surely that meant that A *must* always pair to T, and G *must* always pair to C.

It was the DNA pairing law and once the duo cracked it, they understood in a flash how a living thing copies its genes. Each strand of the DNA double helix was like the positive and negative of a film, each able to reproduce the other. So for instance, a string of A–G–C–T–A–G–C–T would be a template guiding bases to thread into a string of T–C–G–A–T–C–G–A. And so on, ad infinitum, each strand faithfully reproducing a copy of the other.

The Eureka moment occurred on Saturday morning, 28 February 1953. As theatre critic turned science writer Horace Judson wrote in his monumental historiography:[38]

That morning, Watson and Crick knew, although still in mind only, the entire structure: it has emerged from the shadow of billions of years, absolute and simple, and was seen and understood for the first time … A melody for the eye of the intellect with not a note wasted.

Physicists had been looking for a new fundamental law of life, and Watson and Crick had certainly found one. As Crick told everyone within earshot over drinks at the local pub, 'We have discovered the secret of life.'[39]

In newsreels and newspapers and on the covers of magazines, the world was wowed by the infamous nearly two-metre high double helix that Watson and Crick clamped and screwed together on a table in the middle of their Cavendish lab.[40]

Watson and Crick had discovered a fundamental secret of life. They had discovered how DNA, the genetic material, copies itself. But there was another fundamental mystery to be explained. DNA carried the instructions to make an oak tree, a chicken or a Barack Obama.

But how? It was composed of only four different bases, a very limited alphabet for such complex information.

One of those who came to ogle at Watson and Crick's model helix was Sydney Brenner, a young South African scientist working at nearby Oxford University. Just a year before, as a medical student in Johannesburg, Brenner had been intrigued by an essay written by Hungarian computing pioneer John Von Neumann. In his essay Von Neumann had imagined a self-replicating 'automaton'[41] and the 'code tape' it would have to carry to specify how to reproduce itself. There on the table of the Cavendish lab, Brenner saw Von Neumann's 'coding tape'. He realised that the letters of the DNA double helix were not the final information carried by the gene. They were but a code, a code represented *somehow* by the linear sequence of those letters.

As Brenner described it, 'When I saw the DNA structure was also the first time that I recognized the real concept of the genetic code …

How DNA is copied

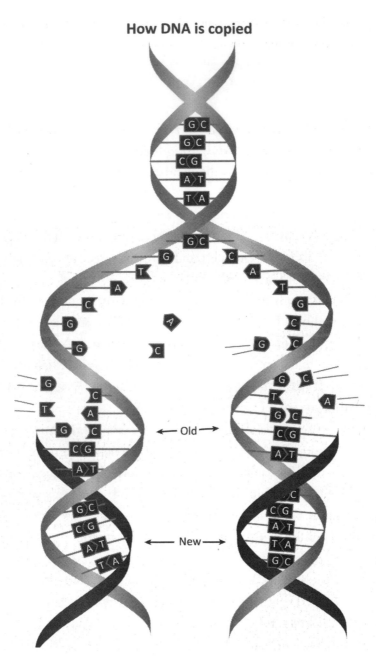

The two strands of the double helix unfurl. Like the negative and positive of a film, each strand provides the template to make a copy of the other.

Biology had been three-dimensional ... But the idea that you could reduce it to one dimension was very powerful'.[42]

***

The double helix revealed a hidden code written in a string of DNA bases. The next quest was to crack this code. No-one knew the cipher but biochemists were pretty sure what the code would translate into: proteins.

Watson and Crick clamped and screwed together this two-metre high model of the double helix. (Courtesy Photolibrary)

Since the 1950s there had been little doubt that genes were required to produce them. Evidence from organisms as distant as funguses and human beings had shown that when a gene was mutated, the consequence was a defective protein. Strafe a bread mould with DNA-damaging X-rays and the mould loses its ability to make an enzyme (a type of protein). Or examine the blood of people who've inherited sickle cell anaemia, and you will find, as Pauling did, that they carry a defective haemoglobin protein. In other words, lose the gene and you lose the protein.

It wasn't hard to accept that proteins should have this role at the centre of life. Proteins are supremely versatile molecules that do all the major work of living cells. For instance, they can be fashioned into enzymes that run the chemistry of life. Like tiny machines, an array of enzymes catalyses the solar-powered reactions that produce sugar from carbon dioxide and water in a plant leaf, while another array of enzymes combusts those sugars to provide energy for the animal that eats the leaf. Or proteins can be fashioned into machines like the motor proteins dynein, troponin and tropomyosin that power the movement of everything from a bacterium's thrashing flagella to a beating heart. Or they can be architectural proteins like actin, collagen and keratin that provide the beams and struts to build cells and bodies.

So if DNA were a code, it had to be a code for protein.

Some thought cracking the code would be easy. DNA was made up of four bases; proteins were made up of 20 amino acids. Cracking the code was a matter of translating from an alphabet of four letters to an alphabet of 20—a trivial problem, thought theoretical physicist George Gamow.[43] It wasn't.

Sydney Brenner provided two crucial clues as to how living cells did it.[44] One was that the DNA of a gene was not directly translated into a string of amino acids. A working copy was made, like a photocopy of an original blueprint. It was called messenger RNA

Decoding DNA

A messenger RNA copy is transcribed from one strand of the DNA template. The RNA letters are decoded in groups of three. Each triplet corresponds to an amino acid.

or mRNA. The copy was of flimsier material than DNA and it differed from DNA in two ways. First, while DNA was double stranded, mRNA was made as a single strand. Second, one of the component bases of mRNA was different: instead of a thymidine (T), it used uracil (U). This disposable mRNA was the working tape that the cell fed into a decoding machine (known as the ribosome) to churn out amino acids.

Brenner together with Francis Crick, now his partner in research at Cambridge, made another key contribution in 1961. They showed that to decode the messenger RNA, it must be read in groups of three letters.

The code was read as triplet

Brenner and Crick deduced that the code was read in triplet based on experiments in bacteriophage. They showed that if one or two extra letters were inserted in DNA it scrambled the code, but that if three letters were inserted the message was recovered. The simplest explanation was that the phase had been restored consistent with the code being read in groups of three. For instance, consider the words CAT DOG RAT. If I introduce a random letter at the beginning, it will destroy the sense of the message: ACA TDO GRA T. Or if I introduce two letters: AAC ATD OGR AT. On the other hand, if I introduce three letters, the message gets back into register: AAA CAT DOG RAT.[45]

But the final cracking of the code was an honour that went to three other scientists. In 1961, Marshall Nirenberg[46] at the National Institutes of Health in Bethesda found that an extract of bacteria could decode mRNA strands. When he added a synthetic strand of mRNA carrying a string of repeating UUUUU he got back a repeating string of the amino acid phenylalanine, showing that the triplet of mRNA letters, UUU, coded for phenylalanine. Using this synthetic approach, Nirenberg, and then Har Gobind Khorana at the University of Wisconsin and Robert W. Holley at the Cornell University went on to decode all the combinations of triplet letters. It turns out that the 64 different combinations ($4 \times 4 \times 4$) coded for only 20 amino acids, because different triplets can code for the same amino acid.

***

The 1950s and 1960s were a grandiose time for molecular biology. From a resonant idea in the mind of Aristotle, the gene had crystallised into the profound form of the DNA double helix. And then its complex message had been decoded from DNA into protein. It had taken the efforts of philosophers, plant breeders, microscopists, fruit fly and maize geneticists, physicists, computation experts and molecular biologists.

Now the 2300-year quest was over. The sense of grandiose finality was captured by Francis Crick's 1957 proclamation of the 'central dogma': DNA makes RNA makes protein. It was a dogma that could hold its own with the greats of physics: the $E = mc^2$ of biology.

And just as the triumphs of physics in the 1930s led many to believe that there was not much left to discover, so too molecular biologists began to feel that all that was left were mopping up operations. 'There was, at that time, a feeling that it was essentially all over,' recalled Brenner.[47] Indeed, Brenner jumped ship to solve the mysteries of animal development in the roundworm *Caenorhabditis elegans*, and Francis Crick headed off to explore the mysteries of the brain.

But the central dogma, though undeniably correct, also fostered a certain hardening of thought and the establishment of an orthodoxy of belief. Over the ensuing years the ramparts of the dogma enlarged, arguably beyond the foundations of supporting evidence. The dogma was enlarged by three tenets.

First, as Crick proclaimed, the genetic code was strictly a one-way affair. 'Once information has passed [from DNA] into proteins it cannot get out again.'[48] It was an emphatic proclamation to blast the vestiges of Lamarckism[49] out of the water. Lamarckist thinking held that life experience could modify genes and explain evolution, the classic example being that giraffes evolved long necks because proto-giraffes kept craning to reach the treetops. If genetic DNA information was a one-way affair, then there was no way for life experience to modify genes. 'So what molecular biology has done, you see, is to prove beyond any doubt but in a totally new way the complete independence of the genetic information from events occurring outside or even inside the cell,' explained French molecular biologist Jacques Monod.[50]

A second enlargement of the dogma was that the genetic code was assumed to be universal.[51] As the effusive Monod remarked in

1954, 'What is true for *E. coli* is true for an elephant.'[52] Monod's statement was triggered by the finding that bacteria would happily decode a gene belonging to a mammal, like part of the rabbit haemoglobin gene.

A third enlargement was that the central dogma came to mean that the *only* information carried by the DNA code was protein recipes. 'Once the central and unique role of proteins is admitted, there seems little point in genes doing anything else,' argued Crick.[53]

Reducing the idea of the gene to a protein recipe was a problem for some people, such as Nobel laureate Barbara McClintock.[54] But she was a mild dissenter compared to Richard Goldshmidt at the University of Berkeley, California, who argued in 1951 'that the gene was merely a figment of the geneticists' imaginations'.[55]

However, for a couple of decades, dissent largely disappeared off the main stage of genetics. How could there be dissent when the central dogma was so spectacularly successful? It underpinned the genetic engineering revolution of the 1970s and launched a multi-billion-dollar biotechnology industry where bacteria were turned into factories for churning out precious rare proteins from human genes, like the protein insulin that keeps diabetics alive or human growth hormone, which rescues children from a life of dwarfism. With factory precision, whatever DNA master tape was supplied to the *E. coli* bacterium, the predicted protein rolled off the production line. The DNA tape was never corrupted by a backflow of information; it was perfectly translated between species, and the *only* information to come out was protein.

Nevertheless, in science, proclaiming a dogma is a perilous thing to do. The central dogma would have to weather many a battering.

In 1970, Howard Temin, Renato Dulbecco and David Baltimore overturned one tenet when they discovered that the code was, in fact, reversible. Many viruses prefer to store their genetic information as RNA. When they take up residence in a genome, they reverse that information back to DNA.[56]

In 1977, the second tenet took a hit when it was discovered that the code is not strictly universal. For anything more complex than bacteria, genes are split by vast tracts of indecipherable DNA.

Like tax legislation, the central dogma survived these challenges by making amendments. It survived this way for 26 years. But the human genome project delivered the greatest assault on the central dogma yet. It fired a broadside at the third tenet: that the *only* function of DNA is to code for protein.

In 2003, as researchers sifted their way through the three billion letters of human DNA code looking for genes, they found that only 1.5% of these DNA letters carried instructions for genes. Within this

The Human Code

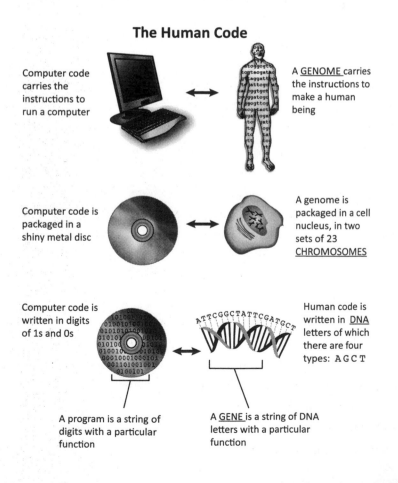

Computer code carries the instructions to run a computer

A GENOME carries the instructions to make a human being

Computer code is packaged in a shiny metal disc

A genome is packaged in a cell nucleus, in two sets of 23 CHROMOSOMES

Computer code is written in digits of 1s and 0s

Human code is written in DNA letters of which there are four types: A G C T

A program is a string of digits with a particular function

A GENE is a string of DNA letters with a particular function tiny chunk, 21,000 genes could be identified.[57] Appallingly, this meant that the noble piece of work that is man is equipped with about the same set of genes as Sydney Brenner's roundworms! If that's the case, then the information for creating a man must rely on some other type of information encoded in the DNA.

As John Mattick, a geneticist at the University of Queensland and one of the most prominent of the iconoclasts put it, 'We may have fundamentally misunderstood the nature of genetic programming in higher organisms.'[58]

2

'Junk is Telling Us Something'

John Mattick delivers a talk like a man in a delirium: fast—an avalanche of words at the speed of thought. Thirty minutes is not enough to sketch out the theory he has built over 17 years. When he is done the audience seems stunned. His theory threatens the central dogma, the very basis of what we've understood a gene to be for the past half century. Yet he gets almost no challengers.[1]

The central dogma states that 'DNA makes RNA makes protein'. Francis Crick, co-discoverer of the DNA double helix, offered the world this terse summary in 1957. A fuller elaboration would read something like this:

> The gene is a recipe for a protein. Each recipe is written in characters called DNA. RNA is a disposable copy of the DNA recipe that directs the production of proteins in the messy kitchen of the cell. It is this final culinary product, the proteins, which are the central actors on the stage of life. They are the ones that do all the work of building cells and running life's chemistry.

So the answer to the secret of a gene is that it is a recipe for a protein. The complete set of gene recipes is our genome, a cookbook for making a human. But ask Mattick for his definition of a gene and you get this:

> It used to be simple: a gene was something that coded for a protein. Now we think a gene can convey lots of information into the

The Central Dogma: DNA makes RNA makes protein

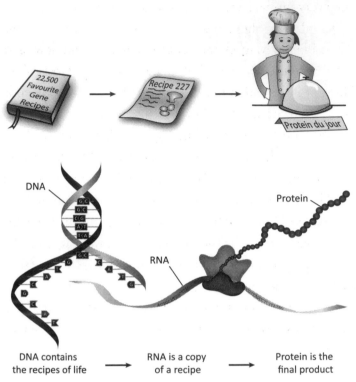

DNA contains the recipes of life → RNA is a copy of a recipe → Protein is the final product

DNA is a recipe book. RNA is a copied page. A protein is the final product.

system. The new view is that the whole genome looks like a continuous gene with all the RNA transcripts interlaced and overlapping. It's almost like we've moved into hyperspace in terms of information coding and transfer.

If you thought you knew what a gene was, and are now totally befuddled, don't worry. You have plenty of company. Many geneticists find Mattick's theory farfetched. Yet Mattick's ideas are making their mark in top scientific journals like *Science* and *Scientific American* and attracting global attention in *The New York Times*.[2]

Mattick came to his heretical view via the digital revolution, the revolution that, among other things, compressed the content of 5,000 long-playing vinyl records into a wafer-thin iPod. Mattick believes something very similar happened to the gene. In the course of evolution it went 'digital'. And the language it used to compress information is RNA. According to Mattick the central dogma should now read: DNA makes RNA. Some of that RNA makes protein. Most of it carries other kinds of information that we are just beginning to decode.

What has strengthened Mattick's hand and made the world sit up and listen are the baffling findings that have emerged since the completion of the human genome project in 2001. It turns out that of our three billion letters of DNA, only 1.5% carry gene recipes for proteins. These gene recipes are of about the same number and type as that of the millimetre-long, 1000-cell roundworm *Caenorhabditis elegans*. We have genes for metabolism, cell building, muscle movement, nerve transmission, hormone signalling, sex. So do they. As Mattick puts it, 'the same parts set'.

So it is to the rest of our genome—the 98.5% that does not code for proteins—that we must look for the answer to what makes us different.

Caenorhabditis elegans. We have the same number of genes as a roundworm. (Courtesy *Science* magazine)

***

Mattick is a born iconoclast. He admits to being 'attracted to the weird', and to having a penchant for 'disturbing the orthodoxy'. He attributes it to the Irish genes inherited from his mother. Mattick can be intimidating. He is tall and solid with dark hair, metal-rimmed glasses, and a greying beard that goes with pushing 60. But it's not his bearing that daunts; it's his forceful, sweeping intellect that hits the mark with articulate, incisive language. The combination of intellect and eloquence is probably part of the explanation for the curious position Mattick occupies. Iconoclasts are often ostracised by their colleagues. Not Mattick. A keynote speaker at international conferences and winner of scientific prizes,[3] he is arguably Australia's most eminent molecular biologist.

His early career gave no hint of a maverick in the making; his scientific pedigree is impeccable. He spent his first 20 years delving into the workings of viruses, bacteria and yeast, and those years of studying how genes work in simple organisms saw no conflict with the central dogma of molecular biology.

Mattick's run in with the central dogma began in a pub in Baylor, Texas, in 1977. He'd just begun as a postdoctoral researcher exploring the genes of the bacterium *Pseudomonas aeruginsoa* at Baylor University and some bizarre news was rocking the scientific community. As we saw in chapter 1, the dogma that a gene was a recipe for a protein was based on what had been learnt from bacteria, but it seemed to be universal. For instance, researchers could supply bacteria with the RNA copy of a rabbit gene and, like short-order chefs, the bacteria would serve up the rabbit protein. In this case they used RNA copies of the rabbit beta globin gene. (Beta globin is one of the subunits of haemoglobin, the protein that makes blood red.)[4]

All well and good, but bear in mind our bacterial short-order chef was using a *copy* of the rabbit gene recipe. Researchers had never spied the original beta globin gene. For good reason. Like a bead on a string, the gene was strung together with thousands of others, bundled up in a chromosome and buried in the dark catacombs of the cell's nucleus. Accessing the *original* beta globin gene would be like trying to find a needle in a haystack. By contrast, accessing the RNA copy was easy. Wily researchers took advantage of the fact that nature had provided a ready source: maturing rabbit red blood cells are dedicated to cooking up haemoglobin and are stuffed with copies of globin RNA recipes.

Researchers took it on faith that the haemoglobin RNA recipes were true copies of the original gene. But taking things on faith can be perilous—as illustrated by this joke.

So there is this order of medieval monks. Their life's work consists of painstakingly transcribing the text of the Bible to spread the good word. One day an earnest young monk reads over one of his copies and realises he has made a mistake. Horrified, he quickly corrects it but a disturbing thought takes hold and he can't shake it off. He seeks out the abbot. 'Father; how can we be sure that the text we are copying from is entirely without error? Could it be that the word we are spreading is false?' The abbot forgives the young monk for his worm of doubt and reassures him that the text he is copying is the unadulterated word of the Lord. The earnest young monk feels unburdened. A few days later the abbot is reported missing. In the midst of alarm and chaos, the young monk has a sudden flash of intuition. He ventures down into the crypt where the precious original copy of the abbey's Bible is stored. Down and down the damp, dark stone spiral staircase he ventures until he hears a faint wailing sound. As he gets closer, he makes out the hunched figure of the abbot rocking back and forth and clutching something large and rectangular to his chest. Between the sobs and mad laughter, he catches the phrase the abbot repeats over and over again: 'celebrate … not celibate'.

Like the abbot, researchers had never actually seen an original copy of a rabbit beta globin gene, or any other mammal's gene for that matter. Until 1977, that is, when they at last found a way to look at the original recipe for the beta globin gene, and like the abbot they got a profound shock.[5] They expected the DNA sequence for the globin gene to be a mirror image of its RNA copy. It wasn't. For starters, it was ten times longer: the original text was interspersed with vast tracts of DNA whose code could *not* be deciphered.

Imagine the DNA code for globin was written with letters. Researchers expected to find the original gene written as GLOBIN. Instead they found something more like **GL**-XXZZZYYY-**O**-QQQQQRRRRR-**BI**-PPPSSS-**N**. What was this gibberish? Was the DNA code not universal after all?

'Once again we are surprised,' was Bob Williamson's reaction in a commentary for *Nature*.[6] Williamson was then a professor at St Mary's Hospital in London, and keenly interested in human globin genes and their heritable disorders, such as thalassemia and sickle cell anaemia.[7] It didn't take researchers long to establish that the globin genes of mice and men were similarly infiltrated with mysterious gibberish.

Mattick was surprised too, and intrigued. Sitting on his bar stool in Texas, mulling over this startling finding, he thought, 'Maybe this is telling us something.' Was there some other kind of information buried in DNA that had nothing to do with proteins?

Surprisingly, mainstream genetics decided that the answer to this question was 'no'. And that was largely thanks to the efforts of Nobel Prize winner Phillip Sharp at Massachusetts Institute of Technology. Sharp showed that whatever gibberish had infiltrated the mother code of the gene in the nucleus, it disappeared from the working RNA copy, so-called messenger RNA, by the time it reached the factory floor of the cell. The gene initially produced a full-length RNA transcript in the nucleus but like an edited home video the gibberish was clipped out and the good bits spliced together again. The bits that were spliced together for export were named exons. The intervening bits of discarded gibberish were named introns. The whole process was dubbed splicing.

With everything neatly named and explained, 'the world collectively breathed a sigh of relief', recalled Mattick. The central dogma had been saved. DNA coded for RNA that coded for proteins. It was just that there was rather a lot of junk in the DNA that was copied into RNA and had to be edited out first. Textbooks canonised the view that for any species more complex than a bacteria, only a tiny percentage of the DNA carried any useful information.

Splicing RNA

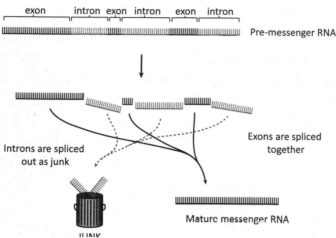

exon   intron exon intron   exon   intron

Pre-messenger RNA

Introns are spliced
out as junk

Exons are spliced
together

Mature messenger RNA

JUNK

Like editing a home video the good bits or 'exons' were spliced together. The intervening bits or 'introns' were considered to be junk.

This relegation of vast tracts of DNA code to the category of 'junk' may seem a little unscientific. But there was a lot of justification. Not only was junk absent of any protein code, about half of it was composed of stutters. A string of DNA letters as short as two or as long as a thousand might repeat up to a million times throughout the genome. How could such inane repetition be a code for anything?

Researchers had a perfectly good explanation for these meaningless tracts of DNA. Part of the problem was that DNA was a molecule that loved to copy itself and sometimes did so quite inappropriately. Another reason was a little more sinister. Our genome bore the relics of past invasions. Eight per cent of our DNA code is actually composed of the corpses of fossil viruses. They were originally retroviruses—specialised viruses that splice their DNA into the host's genes and usually end up killing their host cell—but these fossil retroviruses gave up their marauding ways to become tame stowaways. When it comes to stowaways, the worst offenders are another group known as jumping genes (technically known as transposons.) They account for a whopping 45% of our genome. They literally have the ability to jump around the genome but not beyond the confines of the cell.

Researchers are not sure of their origins. Some may have originally been retroviruses that had their wings clipped; others may be home-grown miscreants.[8] Cells go to great lengths to prevent jumping genes from jumping. But occasionally they do jump and cause havoc by landing inside another gene and scrambling its code. Researchers have linked jumping genes to everything from cancer to Rett syndrome, a form of autism.[9]

Finally, there was another convincing reason to accept that vast tracts of DNA were devoid of information. The total amount of DNA carried by an organism bore little relation to its complexity. A single-celled amoeba, for instance, carries 200 times more DNA than does a human cell. And on it goes: lilies, salamanders and lungfish all have more DNA than we do.[10]

Not everyone dismissed junk DNA.[11] Indeed many were intrigued by these tracts of inscrutable code. In 1968, even before introns were discovered, biophysicist Roy Britten, then at the Carnegie Institution in Washington, found the genomes of complex creatures were littered with repetitive DNA that didn't code for proteins. In 1969, he and developmental biologist Eric Davidson, then at Rockefeller University in New York, proposed that RNA transcribed from this 'junk' DNA might regulate the genes of complex organisms. In the early 1970s, Sydney Brenner at Cambridge University mused on the difference between garbage and junk. Garbage was something you threw away; junk was what you kept in the shed for a rainy day—so junk DNA was likely to be useful over the long term. Physicists like Eugene Stanley at Boston University probed the function of junk DNA by looking for patterns and indeed found long-range interactions more typical of language than gibberish. For Melbourne immu-nologist Malcolm Simons, junk turned to treasure. He took out a patent on junk DNA after finding these sequences were actually a better predictor of tissue compatibility than was the standard test that looked at their human leukocyte antigen (HLA) genes.[12] This junk DNA certainly seemed to be carrying important information.

Mattick wasn't alone in being intrigued by junk DNA but he has been one of the most stalwart soldiers in the campaign to decrypt its meaning.

In the late 1970s, Mattick didn't have a shred of data but he was convinced that introns, together with other large tracts of 'junk' DNA, were 'telling us something'. Evolution would not have put up with junk for hundreds of millions of years if it had no value. One fact in particular played on his mind: only complex cells or eukaryotes, which appeared on the scene 1.7 billion years ago, carried junk DNA. For two billion years before they appeared, life seemed stuck at the single-celled stage, represented by the kingdoms of bacteria and archaea.[13] With the appearance of the junk DNA–laden eukaryotic cell, three complex new kingdoms were spawned: animals, plants and fungi. Could this junk be the key to evolving complexity? The thought insinuated itself under Mattick's skin like an intellectual itch.

***

Scientists tend to understand biology in terms of the technology of the day. Take the brain. The Greeks and Romans viewed it as an aqueduct for cooling blood. Nineteenth-century folk viewed it as a telephone exchange; 20th-century mavens as a personal computer. The netsurfing generation of the 21st century says the brain is like the internet.[14] In Mattick's case, his epiphany about the gene was inspired by the digital revolution.

If I were writing this book 30 years ago, my study would be filled with analogue devices. I would be pecking away on a clunky unforgiving typewriter, doing occasional calculations on a maddening slide rule, and for a well-deserved break heading over to my record player and stack of vinyl LPs. I'd fondle the glossy cover of Debussy's

Greatest Hits, pull out the dinner plate-sized black disc and with all the delicacy my clumsy fingers could muster, guide the playing needle to alight on my favourite track, 'Clair de Lune'.

Now I am pecking away on a forgiving laptop; when necessary I use an obliging pocket calculator; and when ready to relax, I avail myself of the services of a miraculous wafer-sized hot pink iPod. The equivalent of 5,000 LPs is housed within its vanishingly small dimensions yet with invisible nimble fingers my iPod can pick out 'Clair de Lune' in less than a second. That ability to store huge amounts of data and access it so nimbly required a new way to code information. That's what the digital revolution delivered.

Before the digital revolution, all our devices were 'analogue'. They represented information *analogously* to the way it exists in the real world—as a continuum. Length, speed and sound, for instance, were represented continuously by rulers, dials or vinyl records. But there is another way to represent measurements—digitally. Digital representations don't show the continuum, only discrete digits.

A slide rule is an example of an analogue computer. The one I borrowed is made of three slim plastic white rulers that slide past each other. Each is graduated with logarithmic measures so as you add the logs you are actually multiplying large numbers. Seventies maths nerds or engineers like my husband tend to fondle slide rules lovingly and say they are quick and fun to use. Personally, I can barely see the graduations and the slide rule lesson he gave me (I'd never actually used one) was a big headache. And I was shocked to discover how inaccurate they were. Multiplying 23 times 75 (with close tuition) gave me the answer of 1,720. My pocket calculator told me the answer was actually 1,725! The accuracy of the slide rule is limited by the width of the graduations—you simply can't see them if they get any smaller. But apparently this level of error is quite acceptable for the quick reckoning an engineer might have to do.

There are no graduated rulers inside a pocket calculator. It is a digital computer. It does not represent numbers by a continuum but through a code of the digits 1s and 0s. As long as computers of the slide-rule variety represented information with analogue devices, their limited speed and precision put a ceiling on how complex their calculations could be. Once information was represented in a code of digits, it unleashed the digital revolution. Slide rules, vinyl records and typewriters were buried in the dust. Tiny electronic calculators, iPods and laptops grow more powerful by the minute.

In the shift from analogue to digital, Mattick saw a metaphor of the evolution of complex life. The genomes of bacteria are wall-to-wall protein codes: no introns and no junk DNA. Like slide rules, all their operations are transacted by analogue devices—the proteins. By contrast, the genomes of eukaryotes are dominated by junk DNA and introns whose sole product is RNA. Perhaps like digital code, this RNA was a more compact and agile way of representing information.

In other words, to evolve greater complexity computers went digital, and eukaryotic cells turned to RNA. Great theory.

***

For over a decade, Mattick's theory of RNA as a new coding system for eukaryotic genes didn't amount to more than an intellectual hobby. He had been busy building his career as a bacterial geneticist and was invited to establish a new research institute at the University of Queensland in Brisbane, which ultimately became the Institute of Molecular Bioscience. It was a Herculean task and by 1993 he deserved a reward. He would take a sabbatical to 'scratch his intellectual itch'. He headed for Cambridge, where James Watson and Francis Crick had solved the structure of DNA forty years before.

Crick's 1957 dogma held that a gene was a code for protein. As far as information content went, RNA was largely irrelevant. At best, it was just a working copy of the protein recipe; at worst, a useless offcut resulting from the gibberish between genes. Mattick's theory was that offcut RNA was not scrap; it was a new type of code with crucial functions for complex organisms. In good Popperian[15] fashion, Mattick perched himself in the Cambridge library to search for evidence to disprove his theory. 'If you're going to overturn a theory, you have to dress it up and down every which way … The critical observations were the ones that would show it was bunk. Then I could just return to my lab and forget about all this stuff'.

Two bits of evidence threatened to snuff his theory. One was a publication that claimed that the RNA of introns was destroyed within seconds. If the RNA of introns was as ephemeral as a puff of smoke, how could it perform any function? Mattick scrutinised the report closely and found that in fact it referred to the lightning speed with which introns were clipped out of the initial RNA message. But as to how long these RNA offcuts then persisted, no-one knew. Perhaps long enough to do something?

The bigger threat came from the puffer fish or *fugu* as it is known to the Japanese. To geneticists, *fugu* is famed for its tiny genome—about an eighth the size of a human's. To the Japanese, the fish is famous for killing off dozens of intrepid gourmands each year, a result of the paralysing tetrodotoxin its flesh retains if not properly prepared. This little fish also threatened to kill off Mattick's theory. The *fugu* seemed to have shrunk its genome by jettisoning its junk DNA. Many a critic had already waved *fugu* in Mattick's face. If the fish could make organs, swim and poison people without any help from junk DNA, then where was Mattick's theory?

In Cambridge, there was no way of avoiding the truth. Just around the corner from Mattick's library perch was the laboratory of Sydney Brenner, one of the legends who had helped crack the genetic code and then moved on to cracking the mysteries of development in the roundworm. Brenner was also a close colleague of Francis Crick. Now Brenner was again breaking new ground by reading the *fugu* genome. With trepidation, Mattick paid Brenner a visit. Brenner confided in the younger heretic. It was true that *fugu* had jettisoned much of its junk DNA. But it still had introns and some were really big. For Mattick that was a 'cup half full' answer: even the ruthless *fugu* could not afford to lose *all* of its introns. It was enough to keep Mattick's theory alive. So far no complex eukaryote had been found that did not have a sizable amount of junk DNA.

Mattick's scouring of the contents of the Cambridge library also turned up the first bit of evidence to support his theory. The fruit fly possessed a cluster of genes known as the bithorax complex that was crucial for laying out its body plan—for instance, making sure that legs grew off the thorax and antennae off the head, and not vice versa. In between these genes were large tracts of DNA that did not code for proteins. Yet, if they were damaged they corrupted the function of the genes.[16] Those junk sequences produced strings of RNA. What function might this RNA have?

✳✳✳

By the time Mattick returned to Brisbane in 1994, the spark of his revolutionary theory had grown into a flame. He penned the first iteration of his manifesto in the journal, *Current Opinions in Genetics and Development*:

> I suggest that introns, once established in eukaryotic genomes, might have explored new genetic space and acquired functions which provided a positive pressure for their expansion. I further suggest that there are now two types of information produced by eukaryotic genes mRNA [messenger RNA] and iRNA [informational RNA]—and that this was a critical step in the development of multicellular organisms.[17]

Doing experiments back home was difficult. He was still heading the Institute of Molecular Biosciences, and his own laboratory within that institute was focused on the genetics of bacteria, exactly the wrong sort of creature in which to investigate 'informational RNA'. So he kept on with his thought experiment. Assuming his theory was correct—that the RNA transcribed from junk DNA carried complex information—what would this theory predict?

Just as the geneticists of the 1950s had turned to computing paradigms to figure out how information might be stored in DNA, Mattick turned to the information technology experts at the University of Queensland for help. They offered him information theory, explaining that, as systems grow more complex, their regulatory requirements mushroom.[18]

Examples from everyday life abound. Compare a biplane to a Boeing 747. The biplane is all wheels, gears and wings. The 747 has them but it also has kilometres of optical fibres to manage its information systems. A 1950s aircraft engineer examining the wreck of a 747 would recognise the wheels, gears and wings but might well describe the kilometres of optical fibres as 'junk'. Or think of a family garage where Dad not only fixes cars, but also manages three other mechanics and the bookkeeper. If business boomed and the enterprise grew to the size of General Motors, things would change. Half his staff and more than half of the salary would be siphoned off into a management pyramid. He would have managers of hardware and software engineering, business development, marketing and finance all reporting to their respective vice-presidents, all reporting in turn to a CEO, who would report to the Board, who now reports to Dad, the chairman of the company. No doubt some people would also describe the executives at the top of the corporate pyramid as junk.

And so to the evolution of complex life. Like Dad's first car business, the bacteria that colonised the planet 3.7 billion years ago ran a simple operation. Proteins sufficed for their needs. Not only were they the enzymes and builders of cells, they also provided all their regulatory requirements. For instance, bacteria like *E. coli* need to change the enzymes they produce in order to digest different sugars. It involves running off the appropriate gene recipe (from its cookbook of 4,000 genes) for the appropriate enzyme, and conversely not wasting their resources by producing recipes from the wrong gene. The way bacteria stop unwanted gene recipes being copied is to block access. They use special barrier proteins as if they were traffic bollards to block access to a particular stretch of DNA.

Like bollards, these barrier proteins are large and cumbersome and the recipe for making them uses up a lot of space in the bacteria's DNA code—at least a few thousand letters of code per bollard. Bacteria had already given over a fifth of their DNA code to these regulatory proteins. A complex eukaryotic cell had to not only regulate enzyme production, but also to build different types of tissues and coordinate the activities of billions of cells that make up a plant or animal body. If it relied solely on proteins to manage these affairs, it would fast run out of DNA, went Mattick's argument.

Enter RNA. Mattick was banking on the RNA spun-off introns to fill that managerial shortfall. Indeed, junk DNA doesn't produce proteins, but it does produce RNA. If that 'non-coding' RNA was being put to work to manage the corporate affairs of complex life forms, then you ought to see a correlation between junk DNA and complexity. Sure enough, there was no relationship between the total amount of DNA and the complexity of an organism—recall an amoeba has 200 times more DNA per cell than we do.[19] But Mattick and Michael Gagen, his computational colleague at the University of Queensland, took a look at how much junk DNA an organism carries as a fraction of its total amount of DNA. They found an encouraging trend: as the complexity of the organism increased, so did the proportion of junk relative to the total DNA. About 50% of the amoeba's DNA is junk; for a roundworm, it is 75%; for a human being, 98.5%.[20]

All very promising, but there was still one big problem. There was no evidence that RNA did anything but act as a photocopied recipe for a protein.

All that was to change. Mattick had spent his time as a theoretician, building a compelling hypothesis from the top down. Meanwhile the data that would help support his theory began slowly percolating upwards from the nooks and crannies of biology labs.

Some of it from pond scum.

***

In the biological world, catalysing chemical reactions was considered the sole preserve of protein enzymes. So when Tom Cech at Boulder University, Colorado, found in 1981 that RNA could do the same job, he had a hard time convincing anyone. Cech eventually won the 1989 Nobel Prize for the discovery, but in 1981 he had no idea he was about to overturn a dogma.[21] He had merely wanted to study how eukaryotic cells produce and edit their messenger RNA tapes. For his laboratory model he picked the protozoan *Tetrahymena thermophila*, one of the denizens of pond scum that with the aid of a microscope you can see careering around a drop of pond water. He also picked a convenient gene that produced copious quantities of messenger RNA. He found that the RNA initially produced from this gene had a large piece of RNA clipped out—an intron—and then the ends were rejoined to form the final product. He expected to find a protein enzyme that did this splicing. But no matter how much the researchers in his lab searched, they couldn't find one. Even when all the proteins were stripped away, the gene still got spliced. The only explanation left was that the RNA of the intron was able to self-splice. Somehow it was able to clip itself out of the main RNA tape and rejoin the ends. This cutting and pasting of molecules was the work of an enzyme, but when Cech described this RNA as an enzyme, he outraged his biochemist colleagues. As he wrote in his Nobel speech, 'our use of the words "catalysis" and "enzyme-like" to describe the phenomenon provoked some … heated reactions'. The RNA that could act like an enzyme was dubbed a ribozyme.

Once researchers knew what to look for, ribozymes started appearing all over the place: in bacteria, plant viruses and humans.[22] Researchers also found it relatively easy to synthesise and engineer ribozymes for different catalytic tasks, so spurring a new field of biotechnology.[23]

Notwithstanding their talents, until the year 2000 ribozymes were considered freaks of nature. But in that year it became crystal clear that ribozymes are no sideshow. They are central to life.[24]

Think of messenger RNA as being something like an old-fashioned audiotape. The sound on a real audio is brought to life by passing it across the heads of the tape-player. The protein recipes on the messenger RNA tape are brought to life by passing it across a machine called a ribosome (as distinct from a ribozyme). Since 1956, the ribosome tape-player was known to be composed of proteins (like every other machine in the cell) but surprisingly there was also a large amount of RNA in there. It was assumed to be a novel kind of RNA that acted like bubble wrap for the working protein parts.[25] After 20 years of trying to use X-ray vision to see inside a ribosome (something many saw as a fool's errand),[26, 27] in 2000 three X-ray crystallographers

A ribosome is like an audio tape player

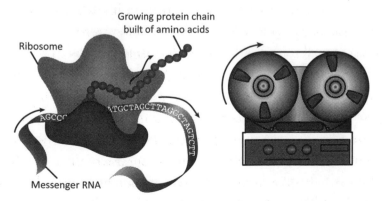

A ribosome reads the information on the messenger RNA tape like an audio player reads sound off an audio tape.

did it. They were staggered by what they found. The working parts of this machine are actually made of RNA; it is the proteins that are the bubble wrap. In other words, a ribosome—the machine that is central to life—is a ribozyme. The three X-ray crystallographers won the 2009 Nobel Prize for chemistry for their efforts.[28]

These revelations about RNA delighted fans of the RNA world theory.[29] This theory holds that life actually began with RNA molecules because they were the candidates most likely to perform the core functions needed for life: catalyse chemical reactions, carry out protein synthesis and replicate themselves. Ribozymes ticked the box for 'chemical reactions'; ribosomes for 'protein synthesis'; and in 2001 researchers showed that RNA could also replicate itself.[30] So RNA now has the proven credentials to be the first molecule of life.

Importantly for Mattick, ribozymes hammered the first nail into the coffin of the central dogma. They showed that a gene could be a recipe for something other than a protein. A gene could be a recipe for a ribozyme.

***

In 1993, Victor Ambrose hammered in a second nail. He had spent 13 tormented years hunting a bizarre gene. In its mutant form, the gene caused a terrible affliction for the poor roundworm, *Caenorhabditis elegans*. A normal worm starts off as a larva and develops into an adult which, being a hermaphrodite, is equipped with both male and female organs. One of these is a vulva from which to lay eggs. However, the poor mutant worm, known as 'bag of worms', fails to develop a vulva. As a result it cannot rid itself of its fertilised eggs; they hatch inside mother worm, devouring her alive. In 1980, Ambrose, then at Massachusetts Institute of Technology, set himself the task of discovering the gene behind this terrible affliction. Part of the reason it took him 13 years to find it was that the gene was vanishingly small. It takes a few thousands letters of DNA to code for an average-sized protein.

Ambrose's hunt ended up narrowing on ever smaller regions of DNA. Ultimately, to his own disbelief, he reported in 1993 that 70 letters of DNA was all that was required to save a worm from becoming a 'bag of worms'. The only RNA transcript to be copied from this DNA was even tinier—a mere 22 letters long. Ambrose was flummoxed. What was he to make of this piece of 'schmutz'[31] as he called it?[32] A mere 22 letters could not code for a protein. What could such a ridiculously tiny piece of RNA be doing that was so crucial for a worm?

It would take seven more years and the efforts of a small army of researchers working on worms, flies, plants and fungi to answer that question. Ambrose had made first contact with a member of a whole new tribe of powerful pygmy genes. These tiny genes were not recipes for proteins. Their active agent was the schmutz, the tiny piece of RNA. Not surprisingly this new tribe of genes were dubbed 'micro-RNA' genes and Ambrose's gene, named lin-4, was the first member to be discovered.

Ambrose got one big clue as to how his lin-4 gene operated. Lin-4 had a curious relationship with a gene called lin-14, a large traditional protein-coding gene. If lin-4 messed up a letter of its tiny code, thus condemning the worm to a terrible fate, that fate could be averted if the lin-14 gene also changed a letter of its code. Somehow these genes were partners in a strange tango. But how did they get together?

Ambrose's colleague Gary Ruvkun at Massachusetts General Hospital happened to be reading the lin-14 DNA sequence. The two scientists got on the phone. They read each other the respective DNA sequences of their genes and were stunned. Buried within the sequence of the big lin-14 gene was a sequence that matched the tiny lin-4 gene. It didn't take them a second to guess what was going on.

RNA, unlike DNA, usually only exists in the cell as a single strand. But if matching pieces were to be made, they would stick together. The schmutz produced by the lin-4 gene would most likely stick to the messenger RNA for the lin-14 gene. Like a bit of dirt on an old-fashioned audio tape, the schmutz might jam the messenger RNA in the heads of the ribosome, preventing the message from being read. In that way lin-4 might serve as an 'off switch' for the lin-14 gene. Ambrose and Ruvkun turned out to be partly right. Lin-4 was indeed an off switch for the lin-14 gene, whose role was to stop a vulva forming before the worm was ready. When lin-4 failed to appear, lin-14 never stopped suppressing the formation of a vulva.

They were also right in guessing that the formation of a bit of double-stranded RNA was the key to the off switch. But it was not just a matter of gumming up the messenger RNA with a schmutzy piece of matching RNA.[33]

Andrew Fire at the Carnegie Institution of Washington and Craig Mello of the University of Massachusetts showed that if you wanted to turn off a gene in a roundworm, the most powerful way to do it was not to introduce a matching piece of single-stranded RNA. The most effective way to trash the message of the gene was to introduce a matching piece of double-stranded RNA. They got the 2006 Nobel Prize for the finding, which was dubbed 'RNA interference'. Yet Fire and Mello had not explained how the double-stranded RNA 'interfered'. It could not gum up a gene message because its sticky bits were effectively glued together like two bits of sticky tape. There had to be some other mechanism by which double-stranded RNA was so deadly to genes with the same DNA sequence. By 2000, the small army of researchers finally clarified what was going on.[34]

The 'other mechanism' was impressive indeed. It was as if the researchers rolled back a dusty roller door and discovered the headquarters of the cell's guided missile defence system. It turns out that all eukaryotic cells[35] are equipped with a hitherto undreamt-of weapons system directed against double-stranded RNA. Double-stranded RNA is seen as a menace. It alerts a special enzyme that dices the double-stranded RNA into bits that are about 20 letters long. These 20-letter bits become nosepiece guides for a missile that cruises off to destroy any RNA messages that sport the same bit of code.

Plant researchers revealed that this seek-and-destroy system was a primitive defence against viruses—most of the viruses that menace plants are made of double-stranded RNA. Since 1929, they'd known that rubbing a small amount of a mild virus on a plant's leaves immunised it against more lethal varieties but had no idea how. In 1997, David Baulcombe at the Sainsbury Laboratory in Norwich, UK, and Peter Waterhouse at CSIRO, Australia, found that the agents of immunity were 20-letter-long pieces of double-stranded RNA that travelled from cell to cell, immunising the entire plant. Now we know these 20 letters bits serve as guides for the missiles that destroy the viruses.[36]

RNA interference — a weapons system

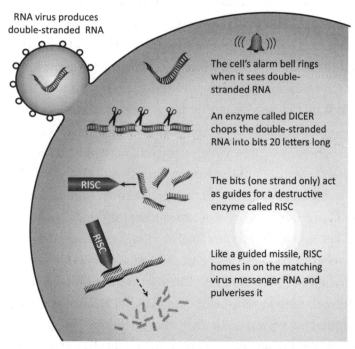

RNA virus produces double-stranded RNA

The cell's alarm bell rings when it sees double-stranded RNA

An enzyme called DICER chops the double-stranded RNA into bits 20 letters long

The bits (one strand only) act as guides for a destructive enzyme called RISC

RISC

Like a guided missile, RISC homes in on the matching virus messenger RNA and pulverises it

A double-stranded RNA virus triggers an attack on the virus messenger RNA. But any double-stranded RNA will draw fire on the matching messenger RNA, a finding being exploited by drug developers to destroy gene recipes.

Elsewhere, researchers showed how the 'seek and destroy' system protected the genomes of roundworms against 'jumping genes'. These restless residents of genomes are harmless as long as they stay put. In order to move they first produce strings of double-stranded RNA. Again, these alert the cell's defence system. Researchers showed that when the defence system was disabled, the jumping genes went on a rampage, causing massive disruptions to the roundworm genome.[37]

The RNA interference system was discovered in plants, fungi, worms and flies, showing that this defence must have evolved in a very ancient common ancestor. Way back in the slime, the first genomes must have found themselves in an existential struggle against parasitic RNA molecules that threatened to corrupt their identities. The RNA interference weapons system was the solution.

In time-honoured tradition, technologies developed under the threat of annihilation find other uses. Swords get hammered into ploughshares; DARPA's defence system for communications gave us the internet.[38] The RNA interference weapons system has also been deployed for peaceful uses: to command and control the activity of genes. And it's not just important for worms. It turns out that human beings have a close relative of Ambrose's 'bag of worms' schmutz gene. In humans it goes by the name of miR 125b-1. And when it is missing, it leads to cancers of the breast, lung, ovary and cervix![39]

The human cancer–microRNA gene connection was first discovered by Carlo Croce, a cancer researcher at Ohio State University. He had spent seven years looking for the genes that drive the most common form of leukaemia (chronic lymphocytic leukaemia). Like Ambrose before him, he'd gotten closer and closer to his quarry only to find it seemingly vanish before his eyes. It turned out that two tiny genes, miR-15 and miR-16, were the culprits.[40] About a thousand of these schmutzy genes have now been discovered in the human genome.[41]

Often these genes lie within the junk DNA of introns, but they can crop up virtually anywhere. Typically they produce a piece of

RNA that folds back on itself like a hairpin.[42] That creates double-stranded RNA: the code red that mobilises the defence force to destroy the message from any gene with matching sequences. Each tiny microRNA gene can control the outputs of hundreds of other genes. Researchers estimate that about two-thirds of all protein coding genes in the human genome are controlled by the microRNA genes.[43]

As Croce found, these controls can be profound. MicroRNA genes seem to play a lead role in cancer.[44] Cancer researchers, who traditionally tested the activity of protein-coding genes to predict the aggressiveness of tumours, now say that the status of microRNA genes has far more predictive power.[45]

And it's not just cancer. MicroRNA genes have also been implicated in age-related macular degeneration, heart disease, Parkinson's disease, Alzheimer's and schizophrenia. These tiny genes are formidably powerful in ways that researchers have yet to fathom. Yet there is an unbelievably fortunate flip side. Researchers can copy the schmutz trick. Just name your gene or virus that you'd like to extinguish and make your own microRNA gene to target it. Researchers are trying to turn off everything from HIV to the rogue jumping genes responsible for causing Huntington's disease. Several drugs based on RNA interference, or RNAi, are in the pipeline including ones to treat HIV, liver cancer, macular degeneration and respiratory syncytial virus infection.[46]

MicroRNA genes break the central dogma because they are not protein recipes. They produce a bit of RNA that, like the feed-forward routine[47] in a piece of computer software code, controls a plethora of other functions.[48]

Between genes that code for ribozymes and genes that code for microRNA, you might have thought Mattick's theory was proven. Case closed, he could hang up his hat and go home.

Wrong.

***

Neither of these RNA revolutions—genes that code for ribozymes or genes that code for microRNA—encompasses Mattick's revolutionary vision. They are, he says, 'bricks in the wall'. But they are not the revolution he is trying to foment. For one thing, only a minuscule percentage of the junk DNA of a human being appears to code for microRNA genes or ribozymes.

In 2000, Mattick at last spied a way that he might gather evidence for his vision of the genome. Tools to explore non-coding RNA—the modern name for RNA produced from junk DNA—had well and truly arrived in the form of heavy-duty sequencing machines and the cost of the technology was plummeting just as surely as the cost of mobile phones or personal computers. Today's 'next generation' sequencing machines can read the entire three billion letters of a human genome in a week for about A\$5,000,[49] a task that took the combined forces of the human genome project 11 years and billions of dollars to complete. Sequencing machines not only read the sequence of DNA letters, they can also read the letters of RNA transcripts.[50]

It was time for Mattick to fully devote himself to discovering the function of non-coding RNA. He began making succession plans for his position as institute director, and winding down the bacterial research projects in his lab. He also struck up a collaboration with the RIKEN institute in Kobe, Japan, which had amassed a vast DNA sequencing factory. Mattick was keen to carry out 'deep' sequencing of RNA. Up until then, researchers had harvested the most abundant transcripts of RNA, which usually coded for protein. Like duckweed floating on the surface of pond, these abundant RNA strings obscured the view. Deep sequencing involved clearing them away and seeing what lay below.

Plenty, as it turns out. The bottom-dwellers turned out to be an astonishing species. For starters, RNA transcripts were not limited to genes and their introns. When researchers looked at different tissues they found that the entire genome was busy making RNA transcripts of various lengths. Could this be a new species of informational RNA?

(On average, a brain or skin cell made RNA from 10–20% of the genome but looking across many different cell types it was clear that the entire genome was open for business.)[51]

The bottom-dwellers were bizarre. Researchers expected to find transcripts that were edited and then spliced together like so much videotape. Furthermore, editing was supposed to be an orderly process that took place within the confines of a single gene: the gene was supposed to produce a long primary transcript, then the unwanted bits (introns) were snipped out and the rest (exons) were rejoined in their original order. And only the strand of DNA that carried a protein code was supposed to produce an RNA transcript; the opposite DNA strand would have produced nonsense.

Not so. Among the population of bottom-dwellers, some showed evidence of bizarre editing. Some had stitched together bits of RNA from different genes, or from non-coding sequence, or from the nonsense strand of DNA. The editing was a complete mess. It was like watching an episode of *Seinfeld* with bits of *Buffy the Vampire Slayer* cut into it, as well as bits of the 5-4-3-2-1 trailer and blank tape sewn in for good measure.

You might say a lot of these crazy edited transcripts are just mistakes by the enzyme machine, called RNA polymerase, that copies RNA from the DNA template. Like a drowsy cameraman, RNA polymerase might occasionally doze off leaving the film running. The editing team splices together the good stuff but occasionally the sticky offcuts left on the floor accidentally stitch themselves together. Perhaps this was the origin of the stuff being trawled by the deep sequencing projects?

Yet many of these crazy transcripts were found over and over again in flies, mice and humans.[52] It was not so easy to dismiss them as rubbish if the same rubbish was being produced in species separated by hundreds of millions years of evolution. When code is preserved across chasms of evolutionary distance it is one of the key indicators of an important function.

In the first seven years after the completion of the human genome project, several exploratory groups dived down to explore the deep sea bottom of the 'transcriptome' and caught a glimpse of the crazy wildlife.[53] Their findings provoked an enormous outcry. Jean-Michel Claverie at the University of Méditerranée School of Medicine— Institute for Structural Biology and Microbiology in Marseilles wondered where it would all end:

> The intergenic, intronic, and antisense transcribed sequences that were once deemed *artifactual* are now a testimony to our collective refusal to depart from an oversimplified gene model. But what if transcription is even more complex? Could it, for instance, lead to mRNAs generated from two different chromosomes. Transcription will never be simple again, but how complex will it get?[54]

Within a couple of years he did indeed find that transcripts could be stitched together from parts grabbed from different chromosomes.[55] Many researchers were bewildered. A common refrain was heard emanating from the journals: What is a gene? 'The idea of genes as beads on a DNA string is fast fading,' wrote Helen Pearson in her news commentary for *Nature*.[56] 'Genes, move over,' began Elizabeth Pennisi in her news piece for *Science*.[57] 'When we started the ENCODE[58] project I had a different view of what a gene was,' said Roderic Guigo at the Center for Genomic Regulation in Barcelona. 'Discrete genes are starting to vanish.'[59] '[This] underscores that we have still not truly answered the question, "What is a gene?"' said Alexandre Reymond, a medical geneticist at the University of Lausanne, Switzerland.[60]

A group of scientists who define genetic terms, known as the Sequence Ontology Consortium, were forced to try for a new definition of a gene. According to Karen Eilbeck, who coordinated the group, it took 25 of them the better part of two days. 'We had several meetings that went on for hours and everyone screamed at each other.' In the end they settled on, 'A locatable region of genomic sequence, corresponding to a unit of inheritance, which is associated with regulatory regions, transcribed regions and/or other functional sequence regions.'[61]

This bewildered chorus was music to the ears of at least one person. For Mattick, it was the anthem for a new order: the revolution had come at last. Ribozymes and microRNA were not odd lost tribes, but the outposts of a vast hidden civilisation populated by RNAs of diverse vocations.

***

Mattick has leapt into the fray to discover the professions of these novel RNAs. It is a vast undertaking. The entire genome is open for business but only a tiny percentage is a code for protein. That leaves an awful lot of RNA sequences to explore.

Researchers like Sean Grimmond, Mattick's colleague at the University of Queensland, are making an inventory. Commenting on the millions of transcripts already accumulated in the FANTOM (Functional Annotation of Mammalian Genomes) consortium database, he acknowledged the challenge of trying to make sense of the deluge of data: '[But] we're getting good at asking questions about ludicrous amounts of data.'[62]

The main questions being asked have to do with where and when these transcripts are produced. In other words, gathering circumstantial evidence. For instance, if a particular transcript is produced in one particular part of the brain at a particular period of its development, then like a suspect at the scene of a crime, it is implicated as playing a role in brain development. It's not proof. However, gradually more and more of the suspects are being nailed, either by adding or subtracting them from cells and looking for an effect.[63]

Observing the antics of RNA is like watching a slum child playing with scraps and producing extraordinary gadgets. It is in equal parts dazzling, undreamt of and totally chaotic. As Jean-Michel Claverie put it, 'Evolution is very eager.'[64]

Let me give you some examples of suspect RNA.[65]

Suspect name: **BC1**	152 letters long and seen loitering in the brain. Mice born without it are anxious and won't explore their environments.
Suspect name: **Xist**	1,700 letters long. Coats one of the two X chromosomes of developing mammals and shuts it down.
Suspect name: **p15AS**	38,400 letters long. Accelerates the growth of cancer
Suspect name: **SCA8**	Causes neurodegenerative diseases like spinocerebellar ataxia
Suspect name: **DHFR upstream**	Uncertain size but like a serpent entwines itself round the DNA double helix and turns on the neighbouring dihydrofolate reductase (DHFR) gene.

How does the scientific community view junk in 2011?

There are some points of agreement. Nobody blithely dismisses junk DNA anymore. Junk DNA produces RNA that is capable of amazing tricks. Most would agree with Claverie when he describes evolution as being 'very eager' to try those tricks out. Like the stuff you keep in the garage for a rainy day, junk can be extremely useful. Complex organisms probably hold on to it because the ability to tinker is a huge advantage to the survival of a species. No complex organism has been able to do without it, not even *fugu*.

Yet while Claverie acknowledges that *some* junk has been put to clever use, he still thinks that most junk is still *junk*, lying around waiting for a function.

Mattick disagrees.

It seems to have come down to a numbers game: How much of junk is actually functional? Mattick and Ewan Birney at the European Bioinformatics Institute have agreed to set the figure at 20%.[66] 'It's a psychological turning point,' believes Mattick. 'Once a fifth is shown to be functional, no-one will call it junk.' And by June 2011, Mattick believed that point had probably been reached. 'I think most people would accept that 10% to 20% looks solid. But Ewan has not capitulated yet!'[67]

Meanwhile Mattick's conviction grows ever stronger. To him, the human genome is an RNA machine.[68] Like Microsoft Word with its densely packed subroutines and feed-forward programs, non-coding RNA is the information-dense language that expands the information in our genome and orchestrates its day-to-day functions. But he is convinced that our genome is way ahead of anything that IT designers have yet imagined. 'The genome is so sophisticated that there are lessons in information storage and transmission that will be really useful for IT once we figure it out. The human genome is about the same size as Microsoft Word, but it makes a human that walks and talks.'

Notwithstanding his successes, Mattick still seems on the fringe. And you get the impression that's just where he likes it.

3

Lamarck Returns

Thomas Henry Huxley, a surgeon and naturalist, went down in history as 'Darwin's bulldog' for being such a ferocious defender of his colleague's theory. He himself was sold on the theory when he read Darwin's *The Origin of Species*. 'How stupid of me not to have thought of that,' was his comment.[1]

Indeed, one of the remarkable things about Darwin's theory of evolution is how obvious it seems when you first hear it. Put simply, parents give rise to offspring who are all slightly different from each other. Those best adapted to their environment thrive, have the most babies, and in the fullness of time give rise to a new species. The idea is so familiar that phrases borrowed from Darwin, such as 'natural selection' and 'survival of the fittest', have become clichés.

Darwin was by no means the first to offer a theory of evolution. The beginning of the 19th century was brimming with theories on the 'mystery of mysteries'.[2] Where did new species come from?[3] The Bible taught that the full repertoire of species was laid down in one act of creation and that was it. But probing the rock strata of England for coal seams showed otherwise. The new science of geology revealed the different layers were ancient sediments, laid down in succession one atop the other like the layers of an extremely ancient archaeological dig. And entombed within the layers were the species of the times. As the rocks changed so did the species. The bottom-most carried extinct shellfish like brachiopods, ammonites and echinoids, the next fossil fish, then monstrous reptiles. The layers were so consistent that William Smith, England's father of geology, could use the fossils like labels to locate coal-bearing seams across the country.[4]

Species were clearly changing over time. But how?

Fifty years before Darwin's theory was published, the French naturalist Jean Baptiste Lamarck proposed what also seemed a rather intuitive idea of evolution, namely, that the experience of individuals—how they used or did not use their bodies—is passed on to their offspring. So, because giraffe ancestors craned their necks to reach the treetops, their offspring were born with longer necks. Or because mole ancestors did not use their eyes, their offspring were eventually born eyeless.

For both theories, the environment was a key factor—but for Darwin it merely selected from differences among offspring that were already there, while for Lamarck the environment or experience actually created those differences. In Lamarck's words, 'An alchemical complexifying force drove organisms up a ladder of complexity.'[5]

Darwin's theory trumped Lamarck's. Any dog breeder could prove Darwin right and Lamarck wrong. If you were trying to breed a lapdog, it didn't matter a hoot how you treated the parents. It only mattered that you consistently selected the smallest dogs from the litter.[6] Furthermore, it didn't matter how many times you cropped the tail of a bulldog, the pups were always born with full-length tails.[7,8]

By 1942, geneticists had hitched Darwin's theory of evolution to Mendel's gene theory. The marriage was formally celebrated in a landmark text written by none other than the grandson of Darwin's bulldog: Julian Huxley. He titled it *Evolution: The Modern Synthesis*.[9] The Modern Synthesis backed Darwinian evolution with provable genetic mechanisms. The raw material of evolution was genes. They varied as a result of mutations caused by radiation or chemicals that zapped genes as randomly as a stray bullet.

Lamarckism by contrast failed pitifully. There was simply no mechanism to explain how the use of muscles or eyes or any experience at all could mould the next generation's genes that were quarantined in eggs or sperm. Lamarckism came to be seen as a scientific joke. Certainly any scientist who dabbled in it tarnished his career.[10]

So, recent reports published in respectable journals might come as a shock. Mice that were intellectually enriched as juveniles gave birth to smarter mice? Water fleas exposed to predators gave birth to babies with protective head shields? Mice fed a folate-rich diet gave birth to healthier offspring, but their offspring's offspring were healthier too? These findings certainly sound very Lamarckian and, indeed, Lamarckism is back. It rode in on the coattails of the genomics revolution. New techniques for analysing DNA are backing up outlandish ideas with incontrovertible evidence. And so many orthodoxies are toppling by the wayside that scientists now dare to tread where once only fools rushed in. Lamarckism also has a respectable new name: epigenetics—literally, something above the genes.

Until recently few people had ever heard of epigenetics. Now even *Time* is writing about it. Because it's not just a phenomenon of mice and water fleas—researchers are finding similar things in human beings. Take, for instance, a study of three generations in the Swedish parish of Överkalix. Birth records going back over a hundred years show that individuals who experienced famine went on to have longer-lived children and grandchildren than peers who did not starve. On the other hand, survivors of the Dutch famine of 1945 seem to be passing down a less enviable health legacy to their offspring. Their children seem to be more obese, more inclined to schizophrenia and, according to one study, more at risk of heart disease. Some researchers are even proposing that the current epidemic of diabetes in China is linked to the starvation of past generations. All of which explains why in January 2010 the cover of *Time* proclaimed, 'Why your DNA isn't your destiny. The new science of epigenetics reveals how the choices you make can change your genes—and those of your kids'.[11]

Epigenetics is the wild frontier of genetics. It is confusing, shocking, heretical, thrilling and laden with hopeful possibilities.

On one hand, it takes us into a brave new world where we have only the foggiest idea of how Lamarckian-style inheritance is taking place. On the other, a loud new chorus of scientists have no doubt it will explain some longstanding mysteries—such as why identical human twins are never really identical; and why modern lifestyles are triggering diabetes, heart disease, cancer, schizophrenia, autism and anxiety disorders. And, possibly, how drugs, diet and lifestyle might reverse those diseases. So, what is this thing called epigenetics?

One of the problems is figuring out exactly what is meant by the term. Randy Jirtle at Duke University in North Carolina likes to use a computing metaphor. If our genome—all three billion letters of DNA code—is like a hard disc, then epigenetics is the program that controls how the information is uploaded.

It's a good metaphor. The problem is there's nothing new about it. Ever since John Gurdon cloned a frog from an intestinal cell in 1958,[12] it's been understood that every cell carries the same (and fully capable) genome. It's just that different cells run different programs—brain cells run the brain program, skin cells the skin program, and so on. So, yes, different types of cells have different epigenetic programs. It's what we used to refer to as 'differentiation and development'.

What's new is that we've never appreciated the extent to which the environment or, indeed, experience, can play a role in programming genes. We thought our programming was mostly hard-wired—directed by an automated program that was built into the genome, as hard-wired as your normal computer boot-up or the way your washing machine runs through its wash cycle. Epigenetics encapsulates a new realisation that the programming of our genes is disturbingly far more 'plastic' than we appreciated.

Sometimes that plasticity can be helpful. Just as Lamarck thought, it can help the individual and their offspring adapt to the environment. So, a besieged water flea mother will grow herself a helmet when she whiffs the scent of voracious fly larvae and her offspring will also be born with helmets. But plasticity can also be harmful. Researchers have been startled to find that epigenetic programming problems may cause human diseases, particularly cancer. Until now, researchers had viewed cancer as a disease resulting from mutations in genes—errors in the hard disc code. Now it turns out many cases of cancer are not problems in the hard disc but a case of faulty programming.

Pondering these environmental effects on a water flea or human disease, it's important to distinguish that we're not talking about traditional environmental mutagens, such as radiation or chemicals. These alter the DNA code. In epigenetics, the DNA code hasn't changed—what's changed is the way it's read. As Jirtle puts it, just as problems with your computer are more likely to involve the software rather than hardware, so most of our diseases are more likely to be epigenetic rather than genetic.[13]

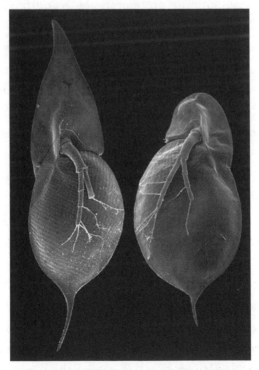

When mother water flea sniffs a predator she grows herself a protective helmet. Her offspring are also born helmeted. (Courtesy Nature Publishing Group)

And when we're talking about environmental factors that affect a foetus, it may also elicit a response of, 'that's nothing new'. We know very well that the developing foetus is highly sensitive to environmental inputs, either because they are mutagenic (i.e. they change the DNA code) or they are otherwise toxic to the precious cells of the embryo. But often it has been hard to nail down exactly what the cause of that toxicity is; even in the case of thalidomide, the notorious morning-sickness drug that created a generation of malformed children, researchers are still not sure.[14] Epigenetics has revealed that some of these environmental effects on the foetus are due to neither mutation nor toxicity but rather to reprogramming.

And let me say from the outset that if epigenetics is the Wild West, there are two districts: the inner zone and the outer limits. The inner zone includes findings on how the environment lays its imprint on genes—startling ones, but still (just) within the framework of traditional genetics. The outer limits are where you find reports that these imprints travel down the generations, just as Lamarck suggested. So far, this is beyond our ability to explain within the framework of traditional genetics.

Most of the findings I'm about to describe belong in the 'inner zone'. I'll warn you when we cross over to the Lamarckian outer limits.

***

Genes are a DNA code but that code has no activity unless it is being 'uploaded' as strings of RNA. Whether or not the information in the gene is uploaded depends on how the DNA has been packaged. If the DNA is loosely packaged, information is uploaded. If it is tightly bundled, it is effectively closed for business. And here is where epigenetics comes in. Something 'above the genes' determines how they will be packaged. It turns out that the 'something' is as unprepossessing as the methyl group—a simple molecule composed of one carbon atom attached to three hydrogen atoms. If the DNA is plastered with methyl groups it will be tightly bundled. (There are also other epigenetic marks that are attached to proteins rather than to DNA: see 'So, how are our genes programmed?', below.)

Bottom line: when code hackers find methyl groups covering parts of a gene's DNA, that usually means the gene is closed for business or, in the jargon, 'silenced'. When the methyl groups come off, that usually means the gene is actively uploading its information.

So, how are our genes programmed?

It's all to do with *packaging*. We've known genes were packaged ever since scientists looked down microscopes in 1902 and saw chromosomes being passed from mother to daughter cells. Chromosomes are DNA packaged in protein.

Packaging is important, but until recently most scientists found it a little boring. Like the masking tape, brown paper and bubble wrap that wraps my latest Amazon book, I rip through it as fast as I can to get to the shiny book cover. So too with the gene packaging; we've known it was important, but compared to reading the text of DNA it was hardly sexy science. The mysteries of epigenetics, however, have suddenly made 'packaging' very interesting.

The DNA of our genome is packed very carefully into the nucleus of each cell. It has to be. Stretched out, it would be a two-metre long gossamer thread—all too easily tangled into a nightmarish knot. The solution, as for any thread, is to use spools. These spools are bead-like proteins called nucleosomes.[15] The DNA thread winds around each bead 1.7 times, runs free for about 50 DNA letters, then winds onto the next bead, and so on, covering about 30 million of them. Spooling shortens the DNA thread to a third of its original length—still too long to fit into a nucleus whose diameter is a few millionths of a metre.[16] To squeeze in, the loosely strung beads start coiling on themselves. And this is where packing starts to get interesting. The coiling density varies from very thick to very light. You can see this for yourself if you look at chromosomes down a microscope: a pattern of darker and lighter bands is visible, corresponding to the denseness of the coils.[17] Densely packed DNA is effectively closed for business. The DNA recipes simply aren't accessible

Packaging DNA

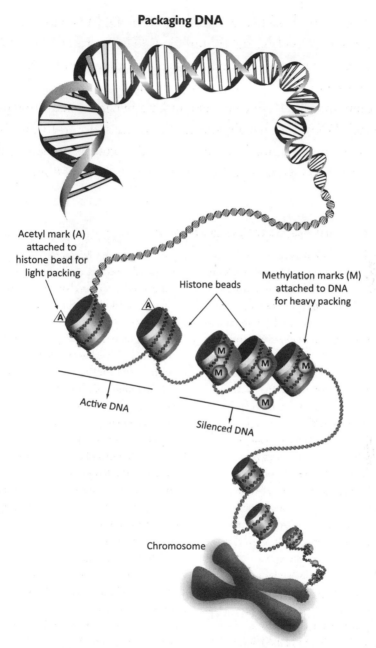

Acetyl mark (A)
attached to
histone bead for
light packing

Histone beads

Methylation marks (M)
attached to DNA
for heavy packing

A

A

M

M

M

M

M

Active DNA

Silenced DNA

Chromosome

DNA is packaged by coiling it around histone beads. When DNA carries methyl groups, it is packed densely, and is silent. When histone beads carry acetyl groups, that makes for a light pack and DNA that is actively read.

to be copied into the RNA that carries out all the multifarious business of genes. Lightly packed DNA, on the other hand, is open for business.

Packing differences explain how different types of cells do different jobs, even though they carry exactly the same set of DNA instructions. In other words, packing is how the cell programs its DNA.

As embryos develop and assign different jobs to cells, their DNA is packed accordingly. Cells destined to form the brain for instance, pack their genes for learning and memory very lightly while their genes for making haemoglobin get mothballed. But in blood cells, it's the other way around. It's rather like home-builders working from an architectural blueprint. The kitchen contractors keep the kitchen plans open but fold away the plans for bedroom and bathroom.

Once a cell has made decisions about how to pack its DNA, those decisions have to be remembered by all of its descendants. It has to be that way or there'd be anarchy—brain cells would divide, forget their packing and turn into blood cells. So every time

a cell divides, not only is the DNA copied, but the particular way that DNA was packaged is remembered.

Packing instructions are provided by chemical tags that are slapped onto the beads like luggage tags.[18] An acetyl group means 'light pack' for active genes. A methyl group means 'dense pack' for inactive genes. These tags can be easily changed, opening up or closing down a gene for business.

There is also an ultra-heavy pack that is harder to reverse. Genes that get this treatment are said to be 'silenced'. Unlike the tags on the protein beads, silencing involves a chemical modification of the DNA itself— a methyl group is attached to the DNA letter cytosine.[19] DNA methylation is the final stage of packing, like the packer from Amazon dabbing masking tape on my book parcel after first having wrapped it in brown paper and plastic. Most researchers track methylation of DNA because it is easier to measure using modern sequencing techniques than the modification to the protein beads.

So now we're ready to begin an in-depth tour. And there's no better introduction to this wild frontier than the Agouti sisters—the rip-roaring poster girls of epigenetics. They are identical twins yet nothing alike. One is a svelte brunette, the other a chubby but still very fetching blonde. It's not because they've dyed their hair differently, been on different diets, spent different amounts of time at the gym, or had different success with their love affairs. The twins are mice. Their couture, lifestyles and love-lives have been identical.

A glimpse of the blonde butterball rocking next to her svelte sister is an arresting sight, especially since the blonde also suffers from diabetes and cancer. Indeed the Agouti sisters—named because the brown one's colouring resembles a South American rodent of the same name—have enthralled geneticists ever since they were discovered. Two ingredients are usually held responsible for shaping an individual—genes and the environment. But for the twins, these ingredients are identical. The mystery factor that made them different was the way one of their genes was read. But this was no ordinary gene: it was a 'jumping' gene. One sister allowed the gene to be read; the other silenced it.[20] Nobody really knows how the mice acquired the jumping gene; perhaps it was through mating with mice from the wrong side of town. There is a well-documented case of jumping genes infecting fruit flies this way.[21] Newly acquired jumping genes tend to wreak havoc in their hosts. In the Agouti mice, the active

The sisters have identical genomes but the brunette has done a better job of silencing her troublesome jumping genes. (Courtesy Dr Jennifer Cropley, Victor Chang Cardiac Research Institute)

jumping gene interferes with a nearby stretch of DNA code that controls appetite and hair pigmentation.[22] The end result of the interference is a fat, blonde mouse. Some of the mice manage to silence the jumping gene by dabbing methyl groups on the offending DNA and so return to being healthy svelte brunettes. Silencing seemed to be a bit of a hit and miss process. In any litter, a mother will always give birth to a spectrum of blonde to brown pups. So researchers had assumed that to silence, or not to silence, was a random event.

It wasn't. Australian geneticist Emma Whitelaw, then at the University of Sydney, made a big splash in 1999 when she reported that in fact the mother's coat colour and weight were heritable.[23] She'd made a closer study of the mice and found that though litters always came in a spectrum of colours, they were skewed to the characteristics of the mother. If mum were a brunette, for instance, her litter would be likely to have 15% more brunettes than if she were a blonde. In terms of numbers it wasn't a big effect, but in terms of scientific impact it was dramatic. As Whitelaw told me, 'This wasn't supposed to happen.' The epigenetic marks were being inherited and researchers had no idea how.

In 2003, Randy Jirtle showed that something else could affect the marks. Supplementing the diets of the pregnant mice with a cocktail of vitamins and micronutrients skewed their litters towards more lean brown pups. The supplements he used—folic acid, choline, vitamin B12 and betaine[24]—were intermediates in the chemical pathway known to drive DNA methylation.[25] And it worked for the pregnant mice. As their pups were developing in utero, the supplements increased the chance of silencing the troublesome jumping gene. It was a fascinating finding, and not just for the mice.

If I may digress for a moment, we've never really known why folate supplements protect human embryos against neural tube defects. These are holes in the brain or spinal cord that develop very early in the pregnancy. Perhaps unruly jumping genes are responsible for these defects and folate helps shut them down? Tempting hints suggest this could be the case. A recent paper linked shifty jumping genes to another human neurological disorder—autism.[26] And another report showed that when pregnant women took folate supplements, one of their babies' genes was coated with more methyl groups.[27] So quite possibly, in humans as in mice, folate supplements may be keeping unruly jumping genes at bay.[28]

But let's get back to the Agouti mice. Scientists had no trouble devising explanations for why the mother's diet helped shut down the jumping genes of her foetus. But they were stumped by the findings of a team of Sydney and Californian researchers. They found that the silencing of the jumping genes was inherited—again, something that was not supposed to happen.[29] A mother rat who received supplements had litters with more brown pups but her pups also went on to have litters with a higher proportion of brown pups whether or not they were given supplements when pregnant. In other words grandma's dietary experience was reaching through the generations to influence not just her offspring but her offspring's offspring. This is pure Lamarckism and classical genetics is hard-pressed to explain it.

For now, the take-home lesson from the Agouti sisters is that diet can, via the plastering of methyl groups, reprogram genes for a lifetime and beyond.

***

Diet can reprogram the genes of other species too (but the reprogramming is not necessarily inherited). Take honey bees. They really are what they eat. Most larvae are fed with honey and become sterile female worker bees who live for about a month. But some lucky larvae receive royal jelly. They become egg-laying queens with a five-year lifespan. No wonder health food shops peddle royal jelly.

So, what's in that jelly? A lot of things, and researchers still aren't sure what the active ingredient is. But Ryszard Maleszka at the Australian National University in Canberra has a fair idea of what it must be doing. When he read the honey bee genome in 2006, he was astonished to find it carried a gene for plastering methyl groups on DNA, much like the gene that sticks them on human DNA.[30] Until then researchers believed insects didn't put methyl groups on their DNA largely because the fruit fly, their favourite insect model, didn't do it. While mammals plaster 60–80% of their genome with methyl groups, in the fruit fly genome they are virtually undetectable. But in the honey bee, some DNA, albeit a paltry 0.7%, is methylated.[31] Why would the honey bee need to methylate any of its DNA, wondered Maleszka? As soon as he asked himself that question he knew the answer. Methylation was a way of producing new DNA programs: surely the honey bee was using methylation to produce different castes of bees! In 2008, members of his lab fed some larvae a substance that interfered with the methylation of their DNA. Lo and behold, they produced queens![32] Something in the royal jelly must be acting the same way. Maleszka suspects it could be phenylbutyrate— a chemical that indirectly prevents the methylation of DNA.[33]

It turns out it's not just bee larvae that change their genetic programming based on their diet. Human foetuses seem to do it too. Epigeneticists like New Zealand's chief science advisor, Sir Peter Gluckman, are convinced that this is all part of an adaptive strategy. Just as a water flea's eggs pick up signals about their future environment and adapt by growing helmets, so too the human foetus is programming itself for the future in response to dietary signals. This 'predictive programming' would work well when the foetal and adult environments match. But according to Gluckman, we are experiencing a great 'mismatch' (also the title of a popular book he has co-authored).[34] We have changed our diet and lifestyle so rapidly that foetal programming is working against us. And therein lies the explanation for the maladies of modern life: obesity, diabetes, heart disease and so on. It's a neat theory but it's far from watertight. And it has to be said, this field is beset by some ferocious disagreements.

The theory has slowly taken shape from pieces of evidence that, like a roughly hewn jigsaw, do not as yet fit nicely together.

The first piece of the jigsaw comes from a tragic natural experiment—the Dutch famine. Human history is replete with famines but this is one of the few to provide enough data to study the long-term effects of malnutrition in the womb. We have to thank the Dutch for their meticulous ways even in the midst of tragedy.

In the closing months of World War II, western Holland was cut off from food supplies. A German blockade had stopped the rail and road services.[35] And then the weather colluded. A severe cold snap froze the canals, blocking the barges. During the five months of famine it's estimated that twenty to thirty thousand people died of starvation. Those who survived did so on a few hundred calories a day, among them thousands of pregnant women. With the defeat of the Germans on 5 May 1945, the blockade ended.[36] Military records show the exact day food supplies returned to the different towns. And in hospitals like Wilhelmina Gasthuis in Amsterdam, babies were born, weighed and their birth records archived. The good people of western Holland did not know that they were collecting data for an experiment that, decades later, would help rock the foundations of genetics.

Some babies showed the effects of starvation straight away. If the mothers starved late in their pregnancy, the babies were born pitifully underweight. On the other hand, mothers who starved only in the early part of their pregnancy went on to have babies of normal weight. But these babies did not escape unscathed. Some were born with neural tube defects (an effect now known to be linked to folate deficiency). For others the effects surfaced much later. In the mid-1960s, the famine babies turned 18 and males were conscripted to the army. A study of the young men showed something quite odd. In an era when obesity was uncommon, they were more likely to be obese.[37] They were also more likely to develop schizophrenia. Subsequent studies probed the Dutch psychiatric registry and found that, indeed,

men and women starved as foetuses were twice as likely to have schizophrenia. It was a finding echoed by studies of the Chinese famine of
1959 to 1961, where 15 to 30 million people starved to death during
Mao's Great Leap Forward.[38] When the Dutch babies reached their
fifties, one group of researchers at the University of Amsterdam took
a look at their risk of heart disease. They reported it was twice as high
as that of their peers.[39] However, here we come to one of the rough
edges of the jigsaw piece. Epidemiologist L.H. Lumey at Columbia
University, New York, has studied the data from the Dutch famine for
20 years and has not been able to repeat the findings on heart disease.
As he told me candidly, when it comes to cardiovascular disease, 'It's
a jungle—there's a whole lot of conflicting data. There's no clear synthesis across the famine studies.'[40]

A second piece of the jigsaw comes from a series of British studies published in the 1980s by University of Southampton epidemiologist David Barker. Combing through 20 years' worth of birth records of
16,000 men and women from the county of Hertfordshire, he found
that babies born weighing less than 2.5 kg went on to develop twice
the rate of heart disease and triple the rate of diabetes compared with
babies weighing in at 4.3 kg.[41] Overall, the lower the birth weight,
the higher the risk of these diseases. His findings were repeated by
other studies in the UK, the US, Sweden and India, and reverberated
through the medical community.[42] The idea that even among pregnant
mothers who were not starving, foetal life experience could influence
the diseases of middle age was alarming. The idea was dubbed 'the
Barker hypothesis', or more colloquially, 'the womb is more important
than the home'.

Barker's studies made a big splash but people didn't really understand what they meant. Was it the final size of the babies that determined their future health or was it some other factor? The Dutch
studies, for instance, showed that normal-sized babies could still
be programmed for disease if their mothers had starved early in the
pregnancy.

Studies in humans could only go so far. So in the late 1980s, Barker made a trip to New Zealand to visit Peter Gluckman at the University of Auckland, who then specialised in studying foetal growth in animals. Gluckman was fascinated by Barker's data, particularly that birth weight and heart disease appeared to show a linear relationship. In a recent interview he told me, 'That's what galvanised me. This suggested we weren't looking at an extreme phenomenon; it was normal adaptive programming.'[43] Gluckman turned to his rats to dissect what was going on. He limited the food intake of pregnant rats and found that their pups, if allowed access to high-calorie food, became obese adults. Compared to normal pups given the same rich food, he says, 'they were doubly fat rats'.

Gluckman seemed to have reproduced the results of the Dutch famine: the starved foetuses were programmed to pile on fat. If rats lived long enough, no doubt they would also develop heart disease. So what was the basis of this programming in the womb? By the early 2000s researchers were becoming aware that there was more to genes than DNA. Along with the new techniques that made it faster and cheaper to read DNA sequences, new techniques were also making it easier to take a look at the methyl groups that were plastered to that DNA. Meanwhile a number of studies in mice and humans were fingering genes that seemed to be involved in obesity. Perhaps some of these genes were being reprogrammed by methylation? Gluckman took an educated guess and tested some of them. Sure enough, in rats starved *in utero*, two of these genes were under-methylated.[44] Could the methylation of these genes also have something to do with foetal programming in human beings?

Gluckman and his University of Southampton colleagues Mark Hanson and Keith Godfrey turned to a 'normal' British population to find out. They managed to coax 550 pregnant mothers to document what they ate during their pregnancy. And when the babies were born, they collected and froze the umbilical cords. Nine years later, they tracked down 78 of the strapping youngsters and subjected them to a thorough medical check-up. They also retrieved their frozen umbilical cords, extracted the DNA and tested particular genes to see if they carried methyl groups. The researchers were startled by what they found. The obesity of the children was strongly linked to methyl marks on one gene—the retinoid X receptor-α gene. The more methyl groups plastering this gene in the cord tissue, the more likely the children were to be obese. Hardly believing their result, the team repeated the study on another 239 mothers and their children at the age of six. The relationship held up. In statistical parlance, the researchers found they could explain a quarter of the difference in the children's obesity according to whether or not the retinoid X receptor-α gene was methylated. How might the methylation of this gene control obesity? The gene is known to regulate the metabolism of fat cells but so far the researchers can only guess at the details of the pathway.

Nevertheless, the results that were published in April 2011[45] caused a lot of excitement. Gluckman described it as 'One of the most exciting data sets I have ever seen. The body fat composition was predictable at birth by epigenetic marks on a single gene.' By comparison he pointed out that no DNA code variations had this predictive power.[46] Jeffrey Craig, a geneticist at the Murdoch Childrens Research Institute in Melbourne, Australia, described it as 'a pivotal finding'. 'This is the first time an epigenetic change detected at birth has been shown to predict a clinically important finding', he told me.

So why did these obesity-prone children have extra methyl marks on one of their genes? Again, the researchers found a startling connection. Mothers who ate a low-carbohydrate, 'Atkins-style' diet during the first trimester of pregnancy were more likely to give birth to babies with extra methyl marks on their retinoid X receptor-α gene. 'It's not uncommon for pregnant mothers in the United States and the United Kingdom to follow an Atkins-style diet,' Keith Godfrey, an epidemiologist and lead author of the study, told me. The researchers believe a low-carbohydrate diet may send a starvation-like signal to the foetuses, putting them out of sync with the high-calorie world into which they are born. Somehow that starvation signal translates into extra methylation of the retinoid X receptor-α gene which predisposes the children to obesity. However, the researchers haven't actually proven that the mothers' diets were responsible for their babies' epigenetic differences; it is possible that underlying differences in the babies' own genes were responsible.[47]

When I offered that argument to Gluckman he countered that, taken together with his rat and sheep studies and the Dutch famine data, it seemed unlikely. 'At some point Ockham's Razor comes to work,' he said, citing the principle that in science one should take the simplest explanation that fits the data.

More hints will come from similar studies Gluckman is about to do in Singapore, where he has been appointed Program Director of the Growth, Development and Metabolism Programme at the Singapore Institute for Clinical Sciences. Singapore has the highest rate of diabetes of any developed country and the rates differ between the ethnic groups. Indians have the highest incidence, followed by Malays, then Chinese.[48] (Indians also have the smallest babies, followed by Malays, then Chinese.)

Meanwhile, back in Holland, the very patient researchers studying the now middle-aged Dutch famine babies also recently took a look at the methylation of their subjects' DNA. Lambert H. Lumey at Columbia University in New York, and colleagues at Leiden University Medical Center zeroed in on a gene that is famous for being regulated by methylation, and for being caught in a battle of the sexes. The gene, called insulin-like growth factor 2 (IGF2), controls the size of the foetus—the greater its activity, the larger the foetus. And here's where the battle of the sexes comes in. The copy of the gene that came from the father's sperm is hyperactive because it is under-methylated, while the copy from the mother's egg is sluggish because it is heavily methylated. It's as if the father is trying to make the baby bigger, while the mother is trying to keep it a manageable size. Between the two copies, the foetus gets the right dose of the IGF2 gene's activity. These parental controls on the gene's methylation were well-known. But now it seems the controls are also finetuned by the mother's diet. Lumey and colleagues found that the IGF2 gene was less methylated in people whose mothers were exposed to famine at the very earliest stage of their pregnancy.[49] Faulty settings on the IGF2 gene might account for their predisposition to obesity and disease in later life. Other studies have shown that people with faulty settings are more at risk of diabetes, heart disease and childhood cancer.[50]

Neither Gluckman nor Lumey are suggesting that the particular genes they have found to be 'over-methylated' or 'under-methylated' are the major players in foetal programming. But like canaries in the coalmine, these genes serve as sensitive indicators for what is going on in the rest of the foetal genome. And they show that dietary factors influence DNA methylation—which is not surprising, as the methyl groups for programming DNA are derived from what we eat. Leafy vegetables are rich in folate, and eggs and meat provide choline and methionine. Researchers have found dietary changes affecting the methylation of genes everywhere from Agouti mice to ageing women.[51]

So it seems you are not just what *you* eat, you are what your mother ate. In Gluckman's view, foetal programming is a way for the foetus to get a head start and adapt to its future environment. When foetal programming matches the adult environment, all is well. A baby gestated by an undernourished mother is born hungry and puts on as much weight as possible to deal with a frugal world. When there is a mismatch and the baby encounters a world of plenty, he or she becomes obese and develops diabetes and heart disease. China's current epidemic of diabetes seems to offer compelling support for the mismatch theory. Nearly a quarter of the population are estimated to have full-blown diabetes or to show the early stages of the disease.[52] Those affected are the offspring of the generation that starved in Mao's Great Leap Forward.

Not everyone agrees with this big picture view. Lumey, who has analysed the Dutch famine studies for 20 years, is more circumspect. 'I'm not ready to spread the gospel of imminent doom based on foetal programming studies; these studies are just not solid enough for such far reaching conclusions,' he told me.

If the data holds up, it suggests that a foetus is so plastic that a mother's dietary quirks, like a 'low-carb' diet, can determine the lifetime course of her baby's health. It could make pregnant mothers paranoid about what they eat. On the other hand, says Gluckman, 'If we can nail this, perhaps we can make a perfect baby … All of this is not so futuristic, we're on the verge of dealing with the evolutionary baggage of plasticity which is no longer working well for us in the modern world.'

***

Peter Gluckman isn't the only one who's convinced that genes undergo finetuning to match the future environment. Moshe Szyf is another true believer. Szyf, now at McGill University in Canada, is one of the early pioneers of epigenetics and I can't help being tickled by his story—because Szyf, an orthodox Jew, seems to have had a philosophical attachment to epigenetics from the very first time he encountered it. 'It's looking for some design in nature. As a scientist, I have to compartmentalise these things but I can't deny there's a religious feeling.'[53] In that vein, Szyf is following in the footsteps of many who have been philosophically drawn to a Lamarckian view of evolution.

Dabbling in Lamarckism is a risky proposition for a scientist. But even Darwin was not completely averse to the idea that the environment could leave an imprint on the hereditary material. The last phrase of his introduction to *The Origin of Species* reads, 'I am convinced that Natural Selection has been the most important, but not the exclusive, means of modification.' Nine years later, he tentatively expounded the theory of 'pangenesis'. Cells of the body threw off 'gemmules' into the blood, which were transmitted to the germ cells[54] that formed the eggs and sperm of the generation. Such gemmules not only carried the instructions for recreating the next generation, but they might also carry a record of the parents' adaptations to their environment.[55]

The 'official' divorce between genes and environment took place after Huxley's Modern Synthesis, which banished Lamarckism from mainstream genetics. This was largely because geneticists could see no mechanism by which normal environmental exposures, such as the food you eat or whether or not you work out at the gym, could leave an imprint on your genes or those of the next generation. Yet throughout the 20th century Lamarckism continued to maintain a powerful philosophical appeal. If individual experience *could* shape the next generation and so contribute to nature's overall design, then that gave life meaning, and one could view evolution as a purposeful, directed process. Darwinism, by contrast, was nihilistic: change occurred not through any directed process but through a random lottery of mutations to genes. Writers like George Bernard Shaw and Arthur Koestler railed against that randomness, and against the ruthlessness of 'survival of the fittest'. They reviled Darwin and defended Lamarck.[56] Koestler even wrote a book to defend the case of Paul Kammerer, a disgraced scientist who had found evidence of Lamarckian inheritance in the midwife toad.[57] Kammerer was accused of scientific fraud and committed suicide. Soviet Russia also embraced Lamarckism over Darwinism. The idea that evolutionary change occurred through an individual's efforts rather than through inborn advantage was a science that bolstered communist philosophy. Like the giraffe striving to reach the treetop, it would be communist ideals, not aristocractic blood, that would improve human character. That is at least part of the reason why Trofim Lysenko, a Lamarckist scientist who focused Russian agriculture on good rearing rather than good breeding, was promoted.[58] Meanwhile, the Darwinian scientist Nikolai Vavilov, famous for discovering the origins of domesticated plant species,[59] was imprisoned and died of malnutrition. Russian agriculture, deprived of improved breeds, went into decline. And 'Lysenkoism' became a cautionary tale of the perils of state-directed science—a tale that was reprised recently when governments with strong religious influences tried to discredit and stifle embryonic stem cell research.[60]

Nevertheless the idea that experience shapes genes retains its philosophical appeal, even for some scientists. Genes and environment were torn asunder in the 1930s because there was no mechanism to explain how they might interact. Moshe Szyf was one of the first to show how that interaction could indeed take place.

***

I confess to being startled when I first saw Moshe Szyf rise from an audience to deliver a talk with a yarmulke fixed to his balding pate. Not too many molecular biologists wear them. He delivered an enthralling talk in his Canadian–Israeli accent, which also provided more clues to his background. Born to orthodox Polish Jews, Szyf lived his first 12 years in many countries but mostly in Iran, where his father was involved in resettling Jewish refugees. By the age of 14 it was time for serious religious study, so he and his mother settled in Israel while his father commuted regularly from Tehran to Tel Aviv. It's hard to believe, but back then the two countries had good relations. Szyf began his studies at the Tel Aviv Yeshiva, a school for studying the classic Jewish texts, then went on to study philosophy and politics at Bar-Ilan University. His parents despaired that he would ever make a living. So he did a 180 degree turn and enrolled in dentistry at the Hebrew University of Jerusalem. 'I am a man of extremes; I went from being a monk to making money. Besides, I wanted to show that people in the humanities were just as smart as those in the sciences.'

The Hebrew University's combined dentistry and medical degree also required Szyf to spend a year in the trenches of a research lab. Once he'd tasted the thrill of discovery there was no turning back. Prophetically, his first encounter exposed him to the methyl group, particularly how bacteria made use of this molecule to alter their DNA.[61] Szyf has been wedded to the methyl group ever since. Indeed, at the end of his talk, among the credits to colleagues was a thank-you to a diminutive pyramid-shaped molecule: one carbon peak atop three hydrogen legs.

Szyf and his colleagues discovered that bacteria pay very close attention to where they place methyl groups on their DNA. So do mammals like us, particularly when they start their embryonic development. The placing and removal of methyl groups was clearly a dynamic thing. At the time no-one quite understood what it all meant.[62] And few cared; this was a backwater of biology. But Szyf remained compelled by the meaning of these shifting methyl marks. Years later as a junior researcher in Phil Leder's group at Harvard and on his way to becoming an immunologist, Szyf jumped at another opportunity to continue his exploration of the methyl group, despite having been advised that it would be far better for his career to just *leave it alone*.

***

It's easy to describe cancer. Normal cells of the body are law-abiding citizens, quietly doing their jobs, respecting their neighbour's boundaries and restraining their growth. Cancer cells are renegades. They quit their jobs, multiply wildly and go on a rampage throughout the body.

It may be easy to describe, but until recently the underlying cause of cancer was a mystery—particularly because it was associated with widely different factors. Radiation exposure causes cancer, as shown by Hiroshima victims. So does cigarette smoking. But cancer could also be caused by a virus and spread like an infectious disease. On the other hand, some people clearly inherit their cancer from a parent. How did all these different factors lead to the same disease?

In the 1970s researchers hit on a grand unifying theory of cancer. Cancer was in fact a disease of the genes, in particular of genes that controlled a cell's growth. They acted like the brakes and accelerator of a car. Oncogenes (literally cancer genes) were the accelerator fuelling the growth of cancers. Tumour suppressor genes, as their name implies, were the brakes. It is these accelerator and brake genes that are the common targets of the diverse factors that trigger cancer. Radiation and tobacco smoke cause mutations that disable the brake genes, or jam the accelerators in the 'on' position. As for viruses, in their travels in and out of animal genomes, they occasionally pick up an accelerator gene. When the virus infects a new cell, the scavenged accelerator gene drives that cell to cancer. When it comes to rare hereditary cancers, the gene that passes down the family line is typically a faulty version of a brake gene. Not only was cancer understood as a disease of the genes, it also provided a case study in evolution. A cell would slowly accrue mutations until it evolved into a species that was 'fitter' than its neighbours. For instance, Bert Vogelstein at Johns Hopkins University in Baltimore, produced a famous 'Vogelgram' showing how a colon cancer went though stages of evolution from benign polyps sprouting on the lining of the colon to increasingly more aggressive and adventurous species. Each stage of the cancer's evolution correlated with the acquisition of different mutant genes.[63]

By the time Szyf landed in Harvard in 1990, the paradigm that cancer evolved through mutations in genes seemed bulletproof. But while there, Szyf learned about a strange case of mouse cancer that did not carry mutations in the usual suspect genes. This particular cancer, an adrenocortical tumour, arose from cells of the adrenal gland. One of the first steps in the evolution of this cancer was the disabling of a brake gene, a gene that produced steroids. But in this case the brake gene carried no detectable mutations. Szyf sniffed the scent of his old friend—the methyl group. Perhaps the gene hadn't mutated; perhaps it had been silenced. It wasn't a theory that was well met. Indeed, his suggestion years earlier to Robert Weinberg (one of the discoverers of oncogenes) that methylation might be involved in cancer, seemed to infuriate the man.[64]

So Szyf did an experiment. He snipped the steroid gene out of the genome of the mouse cancer cell and stitched it into the genome of a bacterium. As the steroid gene multiplied together with the bacterium's genome, its methyl groups were stripped away. Then he replaced this clean gene into the cells of the mouse cancer. The gene worked just fine and produced steroids—at first. But as time went on, the clean gene once again became coated in methyl groups and shut down its production of steroids. It was, Szyf says, 'as if the cancer cell had a program to silence this gene'.[65]

Cancer cells do indeed seem to be programmed to silence genes by methylation.[66] Szyf knew just what to do to try and reverse the situation. A drug known as azacytidine was capable of removing methyl groups from DNA. Researchers had already found that it halted the growth of tumours in mice, though they had never been sure why. Once at McGill University, Szyf founded a company in 1996 called Methyl Gene to develop cancer therapies based on drugs such as azacytidine, which work by changing DNA methylation.[67] One emerging success story is a class of drugs known as a histone deacetylase or HDAC inhibitors. These drugs also keep DNA open and active. They don't target DNA but rather the histone proteins that wrap the DNA. Histones will maintain a loose wrap as long as they carry acetyl tags (small chemicals); the drugs make sure the acetyl tags stay attached. In 2006, the first HDAC inhibitor, named vorinostat, was approved by the FDA for treating cutaneous T-cell lymphoma.[68] Just why it's possible to treat cancer using these 'unsilencing' drugs is still a mystery. You might predict that the result would be a cacophony of genes. It doesn't seem to happen. Szyf imagines that the drugs do something akin to rebooting a glitched computer—allowing cells to find their way back to the correctly programmed settings.

The fact that cancer can be treated by reprogramming genes has put epigenetics firmly on the radar screen. But epigenetics has also won a lot of publicity thanks to another chance discovery by Moshe Szyf.

This one began in a bar in Madrid. Szyf happened to meet psycho-biologist Michael Meaney, a colleague from McGill whom he knew slightly. The psychobiologist and the molecular biologist hadn't had much excuse to chat before. But there over a few beers, they had a most fertile conversation. Meaney was famous for research showing that early life events were crucial in shaping the mental health of a person. For instance, abused or neglected children were at higher risk of depression and anxiety disorders as adults. On the other hand, children who had been raised in nurturing and loving families acquired resilience to life stresses.[69]

Meaney also developed a rat model that echoed these findings. He noticed that some rat mothers were far more nurturing than others. It was even measurable: good mothers spent more time licking their pups than neglectful mothers. Licking made a difference. Pups who'd been well licked grew into confident, adventurous rats, happy to explore new mazes. Those who'd been less licked were nervous and spent most of their time skulking in corners. Even when pups were reared by foster mothers the effects of licking held up, proving that the pup's temperament was dictated by *nurture* rather than nature.

So why were these pups different? Meaney knew the answer. They had different thresholds for responding to stress. A stimulus that barely raised a whisker for the well-licked rat sent a gush of stress hormones surging through the neglected rat. Meaney also knew that there was a particular gene responsible for setting the threshold: the glucocorticoid receptor gene. The more active the gene, the more it dampened the release of stress hormones. The two scientists relaxing in the Madrid bar suddenly found an intriguing hypothesis had sprung up in their midst. Perhaps the glucocorticoid receptor gene was methylated differently in the different rats? By the time the psychologist and molecular biologist left the bar, they had hatched an idea for an experiment.

Back at McGill, Szyf and Meaney tested the glucocorticoid receptor gene in the rat's brain, and they indeed found a correlation between the rat's nurture and the methylation of its glucocorticoid receptor gene. The more nurturing the mother, the less methylated the glucocorticoid receptor gene of her pups. It was almost as if the mother was 'licking away' the methyl groups. And in so doing she was setting the dial for her pup's stress response to 'low'. For Szyf, these findings added a whole new dimension to the meaning of epigenetics. 'It suggested to me that this is no accident. The mother is sending a cue to her offspring to let them know about the world they will live in.'

In other words, rats about to enter a dangerous world would do well to be more anxious, and that anxiety might be instigated by a mother too stressed herself to pay proper attention to her pup.

As a cancer researcher, Szyf had managed to reverse epigenetic changes in cancer tissue using drugs that unsilenced DNA. He and Meaney decided to give it a try in the rats. It worked. If they dosed fearful adults with an HDAC inhibitor called trichostatin A, the rats let go of their anxiety and started exploring their environment. On the other hand, if the drip contained methionine (an amino acid that supplies methyl groups), the rats became fearful.[70] It seems staggering to think that an animal's temperament could be modified by a chemical with such non-specific effects or by a mundane amino acid. Yet, points out David Copolov, a psychiatrist at Melbourne's Mental Health Research Institute, 'many psychiatric drugs are very non-specific'. For instance, most antidepressants work by raising the levels of a very common brain chemical called serotonin.

It's one thing to show a difference between the brains of nurtured versus neglected rats but what does this mean for human beings? Testing the brains of people who were neglected as children is not that easy. Sadly, it's not that difficult either. Over the past few years, Szyf and Meaney have been collaborating with Gustavo Turecki, the administrator of the Quebec Suicide Brain Bank. The bank has a sizable collection of brains from suicide victims with a history of child abuse. The researchers compared the brain tissue of 12 suicide victims who'd suffered child abuse with those of suicide victims who were not known to be abused. Only the child abuse victims showed gene methylation patterns like those seen in the abused rats. Not only was the glucocorticoid receptor gene over-methylated, other genes also changed their methylation settings in a stressed rat-like way.[71] The consistent pattern of change from rats to human beings solidifies Szyf's conviction that he is witnessing a programmed response. 'This is a glimpse at an adaptive genome,' he told me.[72]

***

So far we've dwelt mainly in the inner zone of the epigenetics frontier. The inner zone may be wild but it's manageable; it still fits within the framework of traditional genetics. It includes the terrain we have just covered: the idea that environment or experience can leave its mark on the genes of an individual, be it a water flea, a bee larva, a neglected mouse pup, or a human foetus or child. We don't understand all the controls that determine whether methyl groups will be added or subtracted from genes. But we can glimpse an outline of the general pathway. Diet affects the process because it provides the raw material—the methyl groups. And it's not hard to imagine that other factors such as stress or toxins might interfere with the chemistry that bonds these groups onto DNA. So, from foetal life onwards, diet, stress, exercise, hormones, toxins, a bad boss—you name it—could be doing something to reprogram genes for better or for worse.

Now, however, we head off to the outer limits of epigenetics—the terrain of outright Lamarckism. This is the proposition that reprogramming can last for more than one generation. While there's evidence, there's no explanation that fits within the framework of traditional genetics.

To begin let's go back to the article that graced the cover of *Time* in January 2010. 'The new science of epigenetics reveals how the choices you make can change your genes—and those of your kids,' proclaimed the cover. The story was based on a study of farming communities living in the isolated northern tip of Sweden in the 1800s. The communities were so isolated that they were utterly reliant on what they produced. Some years saw good harvests, some not, and when the harvest failed people went hungry. It's not an unremarkable story except that in the parish of Överkalix, meticulous records of births, deaths and crop harvests survived to modern times. This allowed Lars Bygren and his colleagues at Umeå University, Sweden, to look for links between the nutrition of parents and the health of their offspring. They began with people born in 1905 and went back three generations to 1799.[73]

It is said that if you torture statistics, they eventually speak. Bygren and colleagues did indeed find some tortuous relationships. Men who starved at around 11 years of age produced offspring who lived longer. For women, if they experienced starvation as foetuses, they also produced longer-lived children. The authors explained these correlations by pointing out that this was the time when the father's sperm and the mother's eggs were most susceptible to environmental influences.[74] Among the Överkalix population, the correlation between famine and longevity held true not just for one generation but for two. In other words what the grandparents ate affected the grandchildren's health. Famine in Överkalix seemed to have passed on a health bonus to the offspring—the direct opposite of what was seen in Holland. But perhaps that's because the Överkalix famine offspring—born between 1799 and 1905—were still matched to the frugal world they were born into?

Piquant linkages scratched together from old records are not what you'd call compelling proof of epigenetics or Lamarckism. For one thing, it could be that the offspring of famine survivors were healthier because their parents had better genes in the first place. In other words, plain old Darwinism: the fittest individuals were selected to survive famine and their offspring were likely to be fitter, too.

Nevertheless, Lamarckism did not make the front page of *Time* just because of this one study. Trans-generational inheritance of experience is cropping up all over the place now. It's there in the Agouti mice—our epigenetics poster girls. It's there in *Daphnia*, the water flea. If mother *Daphnia* grows a helmet to protect herself from marauding fly larvae, not only do her eggs hatch out fully helmeted, so too do those of the next generation, even if they themselves lived in predator-free waters. Grandma's lessons are remembered—though not forever. *Daphnia* go back to being helmetless after a few generations in predator-free waters.[75]

Grandparents can pass on some nasty effects, too. Several researchers have found that chemicals that mimic sex hormones, so-called endocrine disrupters, can wreak havoc on the sperm and eggs of mice. These include bisphenol A, a chemical used commonly in plastics, the fungicide vinclozolin and the pesticide methoxyclor. Michael Skinner at Washington State University has found these effects persisted for up to four generations of mice.[76] The mice experiments were carried out with extremely high doses, and the significance of the findings for people remains controversial. But after finding bisphenol A interfered with DNA methylation in his Agouti mice,[77] Randy Jirtle is worried enough to recommend that pregnant women avoid using plastic bottles which may leak bisphenol A.[78]

The final example in this tour comes from Larry Feig's lab at Tufts University in Boston. In 2009, he and colleagues published an experiment that was hard to believe, despite the venerable reputation of the journal that published it. Feig's report showed that mice exposed to a stimulating environment during their youth—a varied collection of toys, exercise wheels and nesting materials—ended up with a superior ability to learn and make memories. That was not so surprising. What was startling was that their offspring were also intellectually superior, even if they didn't get the enrichment. As the authors wrote, 'If a similar phenomenon occurs in humans, the effectiveness of one's memory during adolescence … can be influenced by environmental stimulation experienced by one's mother during her youth.'[79]

Mind-blowing stuff.

So do we have any idea how this Lamarckian inheritance might be working?

***

Catherine Suter has some ideas. She is a dynamic and delightfully genial young scientist at the Victor Chang Institute in Sydney. But like many epigenetics researchers, she's had some tough battles to fight. She and her colleagues were the first to show that what Agouti mice ate affected their grandchildren. They have also found, like Moshe Szyf did, that human cancer can result when methyl groups shut down a protective gene. In their case they found that 15% of patients with human colorectal cancer carried methylated copies of a protective gene called MLH-1.[80] That finding wasn't hard to explain. Mistakes can occur in the methylation process. If a colon cell accidentally ends up with methylated MLH-1 genes, that sets it on the path to colorectal cancer. What's harder to explain is that in some families, the methyl mistake in the colorectal cancer gene was also carried by the children of affected parents.[81] Suter believes this is evidence that epigenetic defects can be passed down the generations. It's a conclusion that is aggressively disputed by other scientists. They argue that the MLH-1 gene may well show a methylation defect but that the underlying cause (and what is being inherited) is an ordinary DNA mutation that Suter and her colleagues have merely failed to detect.[82] As Suter told me, 'There is still a long way to go to convince people.'

It is indeed difficult to convince people that epigenetic marks travel through generations. As Richard Dawkins put it, 'It could only be true if one of our most cherished and successful principles of embryology is overthrown.'[83]

The principle to which Dawkins is referring is that the DNA in our eggs and sperm is believed to be quarantined from the affairs of the body. The principle even has a name: the Weismann barrier.[84] And so, environmental factors might change the methyl marks on the DNA of body cells, but that should have no bearing on the DNA of sperm or egg cells. But what if the Weisman barrier was breached? What if environmental factors could directly change the egg and sperm cells' DNA methyl marks? Even so, traditional thinking held that those changes could *not* be passed on. Because once egg and sperm come together to form the next generation, the methyl marks on their DNA are erased—like someone erasing all the old programs on a computer disc. As the embryo develops, its DNA is overwritten with new programs appropriate to each tissue: heart, brain, skin, bone and so forth. And when the embryo sets aside its own germ cells, later to become eggs or sperm, the DNA of those germ cells is once more scrubbed free of any pre-existing methyl marks. So how could the programming glitches of the parent persist into the next generation?

Suter is one of a growing number of avant-garde geneticists who believe that eggs can carry an inerasable memory of peculiar programming. Like a back-up, that memory is not wiped away when methyl groups are scrubbed off the DNA. So what is this back-up?

Most likely some unusual strings of RNA.

Not so long ago, geneticists considered that RNA's only role was to serve as a disposable copy of a gene's protein recipe. Now we know that it has a powerful role in regulating other genes (see chapter 2, pages 39–49). It may also be the arbiter of Lamarckian inheritance.

The evidence is building that odd little bits of RNA called piRNA can reprogram DNA. They guide a crew of proteins, called PIWI proteins, to plaster methyl groups onto particular genes and shut

Programming DNA

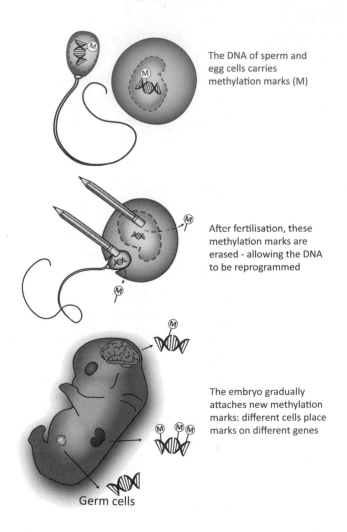

The DNA of sperm and egg cells carries methylation marks (M)

After fertilisation, these methylation marks are erased - allowing the DNA to be reprogrammed

The embryo gradually attaches new methylation marks: different cells place marks on different genes

Germ cells

them down. And they play a crucial function in the germ cells of all animals.[85] They shut down dangerous jumping genes that could scramble the genome.[86]

But what if a regular gene, say a gene for fur colour, accidentally produced a piece of RNA that was recognised by the PIWI crew? The freak RNA would guide the crew back to the fur colour gene and mothball it, just as if it were a jumping gene. As Suter points out,

piRNA keeps jumping genes silent

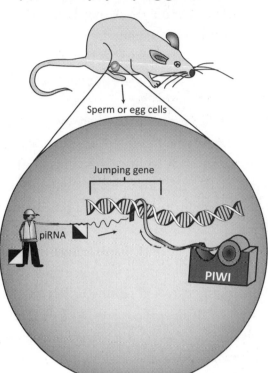

In sperm and egg cells, piRNA guides the PIWI dispenser to tape down jumping genes.

'The system is extremely powerful. Any gene whose RNA accidentally got loaded into PIWI proteins would become powerfully and perhaps permanently suppressed.'[87] The RNA guides might also travel between cells. Like Darwin's gemmules, they might even find their way into egg or sperm cells, causing the gene to be shut down in the next generation. It sounds outlandish but the evidence is building. Over and over in different species, researchers are finding that odd bits of RNA can travel between cells. In roundworms, for instance, researchers can feed them bits of RNA and every cell will shut down the matching gene. Similarly, RNA injected into plants will travel through the cells and circulatory system.[88] And in 2006, a study of mice that inherited a strange spotty-tailed syndrome suggested the effect was caused by itinerant RNA travelling to the germ cells to shut down a gene that affects pigmentation.[89]

Little bits of RNA that can travel around the body and reprogram germ cells sounds eerily like the gemmules Darwin imagined 150 years ago.

Putting these bits and pieces together, we get a picture of how epigenetic changes might be inherited: RNA guides that alter DNA methylation could stow away in the germ cells to affect the DNA of the next generation. But this story has a missing link. Lamarckism, like that seen in the intellectually enriched rats who passed on their cleverness, requires that *experience* leave an indelible imprint on the genes. So is there any evidence that this happens?

Yes. It comes from a weird phenomenon called RNA editing. Not to be confused with the classical editing of RNA to remove introns, this is a process by which an enzyme acts like a director workshopping the script of a play. Just as a director may decide to modify lines, there are so-called 'editor enzymes' that modify bits of the RNA script.[90] It turns out that the brain indulges in some creative workshopping when deciding how to respond to stress. The serotonin receptor regulates the stress response in mice and man. When its RNA is edited, the end result is a wimpy receptor that may transmit a wimpy signal. What triggers the editing? Stress itself. Highly stressed mice trying to find their footing in a water maze edit their receptor to a wimpier form, perhaps to calm themselves down. Clinically depressed people also seem to over-edit their serotonin receptors. They may also be trying to balance their mood, believes neuroscientist Claudia Schmauss at Columbia University, New York, who did these studies.[91]

So these studies show that an experience—stress—can alter RNA. But for Lamarckism to occur, that alteration would have to impress itself on the DNA of genes. Is there a way that altered RNA can change genes? Yes. But we have to look rather far afield for the evidence. Researchers found that in a single-celled denizen of pond scum, *Oxytricha nova*, RNA can overwrite the permanent DNA master copy. From generation to generation, it turns out that it is the RNA recipes of the mother cell that are written back to the DNA master and transmitted to the next generation.[92]

At this point I cannot help but quote Hamlet: 'There are more things in heaven and earth, Horatio, than are dreamt of in your philosophy.' In one place or another in the animal kingdom, the following things are happening: environmental experience is changing RNA. RNA is travelling around the body and entering germ cells. RNA is rewriting DNA code to make its changes permanent. If all this were to happen in the same animal, you'd have all the mechanisms to drive Lamarckian evolution—which is why John Mattick at the University of Queensland recently wrote, 'All of the mechanisms required [for transmitting a memory of adaptation] are either in place or possible.'[93]

So does it happen? It's too early to say whether Lamarckism is a driving force in evolution. So far, in most cases of epigenetics we are seeing a programmed ability to adapt to rapidly changing environments. But it's a flip-flop rather than an enduring change: *Daphnia* bearing helmets eventually go back to being helmet-less when the predators disappear. The species has not changed.

We may have let Lamarck back in on the scene but it doesn't mean we've thrown Darwin out. Nevertheless you never know what is around the corner in the outer limits.

***

Epigenetics remains a wild frontier but it has attracted hordes of glinty-eyed researchers and lots of money. Eager epigeneticists are hoping to see with their own eyes the marks that the environment or disease lay on genes. So now we have the Human Epigenome Project, an international collaboration of researchers equipped with cheap new sequencing machines, embarking on a mind-boggling quest to

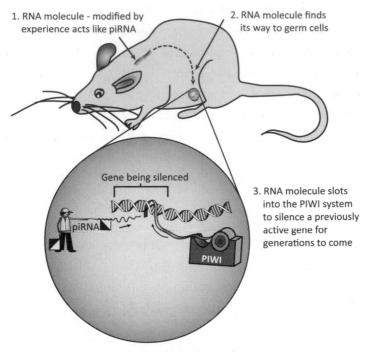

A mechanism for Lamarckism?

1. RNA molecule - modified by experience acts like piRNA

2. RNA molecule finds its way to germ cells

Gene being silenced

piRNA

3. RNA molecule slots into the PIWI system to silence a previously active gene for generations to come

PIWI

All the elements required for Lamarckism to occur have been found to occur.

determine every type of programming glitch our hard disc is capable of. This project will dwarf the genome project in terms of the deluge of data that will flow forth. As Jirtle put it, 'We are going to be doing epigenetics research forever.'[94]

Some researchers think it won't be possible to make sense of the data deluge; they consider it the ultimate in stamp-collecting science: the sort of mindless science that collects data first and worries about what to do with it later. Even Ryszard Maleszka, a devotee of epigenetics in the honey bee, described this global approach to epigenetics as 'funny science'. 'This can't be attacked at a global level; the fluctuations across the epigenome are too dynamic, they change with time and space.' Maleszka is in good company; a number of giants of molecular biology have 'expressed serious reservations about the scientific basis of the Epigenome project'.[95]

Nevertheless, whether epigeneticists are the stamp-collecting types or the more thoughtful variety, the glint in their eye probably goes with the exhilaration of being at a new frontier. They believe that what they find will ultimately reveal the programming errors that underlie modern diseases.[96] And because these are programming errors, they will be amenable to new types of therapies that reverse them, like therapies already being used for cancer.

But a word of warning. Epigenetics is seen as a boon by those in the fad nutrition market. If anyone has got the message that taking high doses of folate or choline to boost DNA methylation is a good idea, think again. In the foetus it looks like too little methylation leads to defects in the nervous system. But many cancers carry over-methylated genes. It's dangerous to go dabbling where you don't understand the controls of the system. Note Robert Waterland's and Randy Jirtle's conclusion to their recent paper '... dietary supplementation, long presumed to be purely beneficial, may have unintended deleterious influences on the establishment of epigenetic gene regulation in humans'.[97]

The glint in the eye of epigeneticists also reflects the thrill of overturning a dogma. Not so long ago, anyone proposing Lamarckian ideas was relegated to the fringes of respectable science. But these days the line between fringe and cutting edge is getting very hard to discern. After 70 years of banishment from serious science, Lamarck is definitely back. And Maleszka for one is not afraid to say it: 'Lamarck was right; the giraffe was just the wrong example.'

4

Your Genetic Future

The opening of the Californian start-up in late 2007 was a carnival scene: gourmet food, music, balloons. In groovy lounges people could deliver some spit,[1] sign away $999 on their credit cards and get a reading of their personal genetic future. The start-up company was '23 and Me'.[2] Its merchandise would help you discover how your 23 chromosomes made you you.

Joe DeRisi was an invited guest. He recounted the scene to me four months later and he was still buzzing.[3] DeRisi is a hard-nosed Californian geneticist who soared to fame in 2003 when his lab helped avert a SARS epidemic by identifying the virus just 24 hours after receiving a sample. I found his titillation with '23 and Me' a little puzzling. And then I began to understand. 'I went to a party and stepped into the future,' he told me. The future that DeRisi and others had predicted two decades ago had suddenly arrived.

This was the future envisaged in the bright eyes of those who had championed the human genome project.[4] In the late 1980s, technological advances had propelled humankind onto an audacious quest for self-knowledge. Forget about being able to inscribe the three and a half million letters of the Bible on a grain of rice; the new techniques were making it possible to read the three billion letters of the human code packaged inside a speck less than a millionth the size of the full stop at the end of this sentence.[5] And once the DNA code of a human being could be read, then it would be just a short step to deciphering why one human being is different from another: fatter, taller, smarter, sportier, has red hair, freckles, loves chocolate or has diabetes, asthma, cancer, heart disease and so on—all the traits and diseases that are common, and that commonly run in our families. Mine has diabetes and heart disease; Kon my neighbour says his family all have high blood pressure; my Tai Chi teacher says the Wongs all have diabetes.

Because these diseases commonly run in families, we've long suspected that there has to be a genetic element. Studies on identical twins raised apart confirmed that. Since they don't share the same environment, their shared characteristics must be due to shared genes.[6] And 40–90% of the time, they do end up with the same diseases. For diabetes and heart disease, multiple sclerosis and blood pressure, genes account for around 40–50%. For autism, schizophrenia and height it's up to 80–90%.[7] These studies on twins get the final say in the nature–nurture debate: it's clear that for most things, genes play a big role.

Like dangling bait the genes that lie behind these common traits and diseases have tantalised geneticists for over 100 years. Starting in the 1980s, they managed to fish out the genes for rare heritable conditions like cystic fibrosis or Huntington's disease, but the genes that explain our garden variety ailments were too slippery to be caught. Those who ventured to read the code of man prophesied their readings would change all that: in the post-genome era, the mystery of what makes you you would be solved. And, of course, right from the beginning people got jittery. The film Gattaca[8] captured the angst. At birth every baby had their genome sequenced and their life's trajectory read out like a weather forecast. Vincent Freeman's forecast was poor. The hero of the story had a 60% chance of a neurological condition, a 42% chance of bipolar disorder, an 89% chance of attention deficit disorder, a 99% chance of a heart attack and a life expectancy of 32 years. He dreamt of a career in space but his mother set him straight: 'The only way you'll see the inside of a space shuttle is

if you're cleaning it.' Vincent tells us, 'It's illegal to discriminate—
genoism it's called—but no-one takes the law seriously.'

***

'Learning about your DNA can help you to understand a little better
why you are the way you are,' reads the advertising patter on the
'23 and Me' website. No wonder DeRisi had that tremulous look
about him: prophecy had just been realised. But not everyone shares
DeRisi's thrall. Companies that offer direct-to-consumer genetic testing are highly controversial. In June 2008, California tried to declare
them illegal; Germany followed a year later.[9, 10] The sudden arrival of
the future has left many people with a bad case of vertigo. Genetic
counsellors, doctors and health advocates are reeling as they witness
the sudden slide into *Gattaca* and contemplate its potential consequences, an era where, as Vincent put it, 'We have discrimination
down to a science.'[11]

There's no doubt that we are heading for the time of complete
genetic readouts at birth. However, companies like '23 and Me' don't
read every letter of your three billion–letter genome; they scan about
one million letters—those that vary most commonly from person
to person—to guess your genetic future. And while the presence of
such companies is sending many people into a tailspin, there is a
somewhat overlooked question that needs to be answered. Can the
tests by '23 and Me' actually predict anything? Geneticists are pretty
unanimous on the answer to this one: no. Not for most common diseases.[12] Despite the proliferation of these companies, the predictive power of their product is puny. For instance, though we know
that genes account for about half your chance of developing diabetes, gene scans can't seem to find more than 5–10% of the culprits.
What use is a genetic test if most of the relevant genes can't be tested
for?[13] Oxford University geneticist Mark McCarthy has spent the
past ten years unearthing the diabetes genes that are currently used

for genetic tests. He says these tests are far less reliable than standard medical predictors: 'A genetic test will be right about 60% of the time [which is only slightly better than 50/50 odds]; body mass index, family history and age give the right answer about 75% of the time.'[14]

It seems we are not at *Gattaca* yet.

But what about the future? Is it only a matter of time before gene scans reveal all there is to be known about a person's genetic destiny?

Curiously, a decade on from the end of the human genome project, we don't know the answer to that question—because there is another fundamental question that we haven't yet resolved. How are common human traits put together? British geneticist Ron A. Fisher had a stab at it early last century. He decided that many genes must be at work. But he couldn't say whether it was hundreds or a handful. If it is a handful, predicting genetic futures should be possible. If it is hundreds or thousands, that could get problematic. One hundred years after Fisher and just over ten years since the end of the genome project, we don't yet know the answer to how common traits are put together—because we haven't been able to find the genes.

***

That failure has led to a minor war between the geneticists. Since the end of the Human Genome Project, some of them spent many years and hundreds of millions of dollars mining genomes to find the genes behind common traits. They came back with little to show. And since 2008, the critics have been ruthless. The stoushes spilt out of hallowed journals like the *New England Journal of Medicine* and into *The New York Times* and tabloids.[15] In 2008, David Goldstein, a geneticist at Duke University wrote in the *The New York Times*:

> It's an astounding thing that we have cracked open the human genome and can look at the entire complement of common genetic variants, and what do we find? Almost nothing. That is absolutely beyond belief.[16]

In 2009, the *Telegraph* quoted geneticist Steve Jones of University College, London, saying:

> It may be time to stop throwing good money after bad. Genetics has been a series of revolutions of diminished expectations … We have wandered into a blind alley and it might be better that we come out of it and start again.[17]

Joseph Terwilliger, a statistical geneticist from Columbia University in New York, argued vociferously that it was high time to abandon the dig. For him, the findings confirmed just what he and his colleague Ken Weiss at Pennsylvania State University had predicted the gene miners would find. When it came to the genes behind common diseases, they were convinced that there were no nuggets to be found. Rather, there would be hundreds or thousands of tiny ones, a mountain of glittering specks of gold dust, largely valueless.

Not surprisingly, the gene hunters were defensive. As Eric Lander, a lead player in the Human Genome Project and an adviser on science and technology to President Barack Obama told me, 'Prediction? It's the wrong question. We're not in this game to find out who will get diabetes. If we never get to the point of prediction but find there are 112 proteins and 17 pathways that cause the disease, then we can fashion therapies, and that's really good.'[18]

The stoush between the geneticists is nothing new. I witnessed an earlier skirmish at a genetics conference in tropical Cairns in 2001 in Australia's far north.[19] The Human Genome Project had just been completed and the field was swept by a rush of gold fever—the human genome was open for mining. Optimistic predictions were running hot. Charles Cantor, CEO of the gene mining company Sequenom, had predicted that all the common disease genes would be nailed by the end of the year.[20] By the time of the Cairns conference a golden speck or two had been unearthed—a suspect gene for Crohn's disease, another for macular degeneration—and the researchers' blood was high. Joseph Terwilliger, one of the invited speakers, seemed like a sheriff trying to hold the throng at bay. Common diseases, he coun-selled, are nothing like Huntington's disease or cystic fibrosis. These so-called Mendelian diseases are caused by a mutation in a single gene. Like a car manufacturer's recall on a faulty production batch, the gene behind Huntington's disease or cystic fibrosis is the same in every case. That gives both the car manufacturer and the geneticist a good chance of nailing the faulty component. But complex diseases like diabetes would be quite different. Many factors could cause the disease and they would differ from person to person. Instead of a manu-facturer's recall, this would be like trying to come up with a common cause of all flat batteries. Any weary mechanic will tell you that there isn't one: a flat battery could be due to a miswired lead, cracked plugs, a faulty door light, or even a boot that doesn't close properly.

In Terwilliger's view, the prospectors were chasing after fool's gold. He said as much in a piece he wrote with his colleague Ken Weiss in 2000 in the journal *Nature Genetics*. It was prefaced with the follow-ing excerpt from William Faulkner's novel, *The Sound and the Fury*, describing the Californian gold rush:

> They all talked at once, their voices insistent and contradictory and
> impatient, making of unreality a possibility then a probability, then
> an incontrovertible fact, as people will when their desires become
> words.[21]

***

For over a decade now, gene miners have searched for disease-causing genes by comparing the DNA codes of healthy people with those of disease carriers. But they haven't checked all six billion letters of the code.[22] They tested up to a million letters, located at those spots where the code tended to vary. The idea was that these common code vari-ations would correspond to common diseases. That approach did not yield much.

But the story is far from over. As I pen these words in April 2011, the tsunami of DNA technology has swept the old arguments aside. The cost of reading all six billion letters of a human genome has plummeted, so much so that it has left Moore's law—the gold standard for dazzlingly fast innovation—in the dust.[23] In 2007, James Watson, co-discoverer of the structure of DNA, had his genome read for around a million US dollars. In 2011, it can be done for as little as US$2500 to US$7000.[24] Around the world, gene hunters are starting to sequence the entire genomes of thousands of people with different diseases and comparing them with the genomes of thousands of healthy people.

It's been a game changer. Arguments about the true nature of the genes behind complex traits will soon come to an end. We will, as McCarthy says, be able to read the genomes of the sick and the well, from top to toe. It will tell us whether complex human traits are built from gold dust or from nuggets. And once we know how we are built, genetically speaking, will *Gattaca* then be upon us? For now, no-one is sure. But as Mark McCarthy told me, 'We should know either way in about two to three years.'

Not bad going, moving from a science fiction movie to the edge of reality in a mere 14 years. That breakneck journey is a tale worth telling. Especially to genome jocks like my son, who sits at a computer screen and searches for genes in a human genome as easily as he plays the latest version of the game 'Might and Magic'. No doubt it's hard for him to conceive what it was like in the dark ages—a mere 30 years ago—when scientists had no idea how to probe a genome.

***

The human genome. What sort of visual imagery does that conjure up? Perhaps endless strings of letters on a computer screen, or perhaps the iconic double helix with the letters forming the rungs of the twisted ladder.

But these are manmade representations. The real human genome is much messier to deal with. It is a gossamer-thin string of DNA letters two metres long. It would be utterly unmanageable were it not spooled around proteins, cut, and packaged in two sets of 23 rod-shaped chromosomes.

As far as visual imagery goes, what the human genome conjures for me is the cutting room in the studios of ABC Radio in Melbourne, over 15 years ago. There I am, barely able to contain my panic as a mound of tangled shiny brown audiotape grows at my feet. I'm editing my interviews by running the reel through the heads of the audio machine and marking the bits I want to cut out with an oily white pencil. I cut at the white marks with a razor and rejoin the ends with tape. Sometimes I make a mistake. I search the mound for the bits of tape I've edited out. But as I go along, the pile on the floor grows bigger and bigger … The precious snippets disappear into it and I've no way of recognising them.

A tangle of featureless tape is largely what the gene hunters had to face. Like the pile at my feet, gene hunters had no way of reading the information in the threads of DNA. Somewhere in the vast expanse, the genes that made a human being lay buried in an ocean of senseless code that used to be called 'junk DNA'. If I got really desperate with my tangle of tape, all I had to do was to run it past the heads of the tape player to hear what was on it. But what could the early gene hunters do to find the genes in the DNA tangle?

What the early gene hunters did was to step their way backwards from proteins. They knew their central dogma: DNA codes for RNA, which codes for the string of amino acids that makes a protein. And they had the cipher. It was just a matter of reading the message backwards from the string of amino acids to get the DNA code.

By 1977, geneticists had their first success. They were hunting the gene for the hereditary blood disease sickle cell anaemia. In this case, nature had provided the gene hunters with a whopping clue.

Clogging up the veins of patients were some brittle, sickle-shaped, red blood cells. Their distorted shape was attributed to crystalline rods of haemoglobin that, like tent poles, had elongated the normally round cells. Haemoglobin was usually a soluble blobby protein. Cleary this abnormal haemoglobin was the cause of the disease.[25] Researchers analysed part of the haemoglobin amino acid code, worked out what its corresponding DNA code must be and synthesised a small section of it. Similar DNA sequences stick together, so they could use this DNA snippet like a lure to fish out the entire gene buried in the tangle of genome DNA. They successfully snared the haemoglobin gene and read its entire 1,600-letter DNA code. They also snared and read the code of a haemoglobin gene from a person with sickle cell anaemia. And they found the cause of sickle cell anemia: a single letter of the code had been changed![26]

The sleuthing process from protein suspect to gene culprit also led gene hunters to the genes for haemophilia, phenylketonuria, and later for hereditary Alzheimer's disease. But these cases were the exception rather than the rule. Most often, researchers had no idea as to which protein had broken down to cause the disease. And without a protein suspect, there was no trail to guide them back to the gene.

***

So, geneticists were stuck. They knew that the threads of DNA carried the code that made a human being. And that ever so slight variation in that code led to the extraordinary variation of human beings: Einstein's brain, Joan Sutherland's voice, Usain Bolt's lightning speed[27] or less enviable inheritances like Michael J. Fox's Parkinson's disease and Halle Berry's diabetes. Being able to read this information was the Holy Grail of genetics. But all was shrouded in mystery. DNA was as unreadable as the shiny brown tangle of audiotape on the cutting room floor.

Things took a leap ahead in 1980. Long before anyone had dreamt up the human genome project, a couple of geneticists thought of an ingenious way to read the human genome.[28] They borrowed an enzyme made by bacteria to act as their reader. So-called restriction enzymes recognise a specific sequence of six to eight letters in the DNA thread and when they do, they cut it.[29] Like the white pencil marks I drew on my audio tape, the restriction enzymes transformed the featureless thread of DNA into a tape with marked sites on it. The geneticists had another crucial insight. If human DNA varies slightly from person to person, then maybe some of the cut marks would also differ slightly from person to person. If we imagine our DNA as a tape with marks one to 100 on it, then some people might be missing mark 45, while others have an extra one, say a 99.5. The cut marks did indeed vary slightly from person to person. For the first time, geneticists had landmarks on the once featureless landscape of human DNA.[30] Now they were ready to ask the six million dollar question: Were some of these variable markers linked to the DNA alterations that cause disease? It was an eminently testable hypothesis. And one of the first places they tested it was with families afflicted by the tragedy of Huntington's disease, a brain-destroying disease caused by a single dominant gene.[31] In these families, if one parent carried the gene, half of the children were affected. Jerky uncontrolled movements usually began in middle age and brain function gradually deteriorated, often ending in dementia.[32] Researchers probed the DNA of affected family members to see if they carried an altered DNA marker. They did. Affected family members often inherited an altered DNA marker on chromosome 4. That meant one of two things: the altered DNA surrounding the marker was itself the cause of the disease, or more likely, the marker lay very close to the gene that caused the disease—so close that the shuffling of chromosome chunks that occurs in families had failed to separate the marker from the disease gene.

Mind you, it took ten years to zero in on the gene. The marker was the beacon that guided researchers into the right part of chromosome 4, but there was a lot of wild territory to explore around the beacon. Imagine being helicoptered to a promising geological site in Alaska and being told to find some gold mines. There'd be a lot of soil you'd have to turn over. That was what it was like for the gene hunters. They had to start digging around their beacon, one shovelful at a time, looking for hints of genes. And then if they thought they had found one, they compared the sequence of the gene in family members with and without the disease. This painstaking stepping across the wilds of the genomic landscape, reading bits of DNA sequence as you go, was called 'chromosome walking'. Mercifully, the days of having to dig shovelful by shovelful are over. The entire terrain has been dug over, or rather sequenced letter by letter, to give us a 'map' of the genome. That was the Herculean achievement of the Human Genome Project. It cost close to $3 billion and took thousands of people and machines 11 years. Nowadays, you still follow a beacon (the variable marker) to bring you into the target zone, but you have a complete map of the genes in the vicinity and what their normal sequence should be.

Chromosome walking may have been painstaking, but the decade of the 1980s was a golden era. By 1985, some five hundred genes for Mendelian diseases were cloned—meaning that they had been located on the chromosome and fished out for further study.[33] Fishing out the genes shone a dazzling light on the cause of these mysterious maladies. Huntington's disease resulted from a stutter in the code for a brain protein—hundreds of repeats of the letters CAG, and the more repeats, the more aggressive the disease. Cystic fibrosis resulted from a fault in a gene whose protein ferried chloride ions in and out of cells. The imbalance of chloride ions led to the gluggy mucus that attracted infection and made breathing a struggle. Hereditary Alzheimer's disease was a fault in a gene for another brain protein (amyloid precursor protein); it failed to undergo its normal pruning and instead got tangled up in toxic dumps. These are just a few examples of 2,565 genes that have now revealed compelling insights into the nature of Mendelian diseases.[34] These genes provided new leads on how to treat diseases—and some have already led to blockbuster drugs such as statins to prevent heart disease.[35] Many more are in the pipeline, like drugs or antibodies that try to reduce the levels of the Alzheimer's protein, and there's light on the horizon for a drug that may fix the faults caused by the cystic fibrosis gene.[36] The greatest value of these gene discoveries so far is that embryos can be screened to see if they carry the faulty gene, allowing parents to prevent a lethal genetic endowment being passed on to their children.

But for all the triumph, the fact is that these Mendelian maladies affect only a tiny proportion of the population. Most of the population is afflicted with illnesses that are not traditionally thought of as 'genetic'. Yet diabetes, heart disease, high blood pressure, cancer and schizophrenia do seem to run in families. So could the genes that predispose people to these common illnesses be found? The DNA markers that had helped locate genes in families affected by Mendelian diseases were no use in tracking down the genes for families affected by these common diseases. Geneticists suspected the reason was that their quarry was not a single gene but many genes colluding to create the illness. And they had no way to find them.

***

The Human Genome Project changed all that. As computers started reading the human DNA code, geneticists were able to see the degree of human variation firsthand. Reading the DNA of different people, it emerged that the code varied about once every thousand letters (we've since revised that to once every 300 letters).[37] At particular addresses on the genome highway, it seemed the code could be swapped to a

SNPs explain the difference

SNPs are single letter changes in the DNA code that largely explain why one person is different from another.

different letter without catastrophic effects (unlike the single letter swap that causes sickle cell anaemia). An A might be swapped for a T, a G for a C.[38] These letter swaps were referred to as single nucleotide polymorphisms or SNPs. Soon everyone just started calling them 'snips'.

Cute and profound. These little SNPs were what the dreamers of the genome project had in mind: they spelt out the reason why one person is smarter, more musically gifted or more athletic than another, and also why they sicken in one way rather than another. These SNPs would allow geneticists to move beyond individual families to gather huge amounts of data from entire populations, reading their SNPs to find the coding errors behind human diseases.[39] Drug companies quickly cottoned on. Sequenom developed a technology to rapidly 'type' the particular collection of SNP variants a person carried. By 2001, Sequenom's CEO, Charles Cantor, was making his prediction that all the disease genes would be unearthed by the end of that year. [40]

And so the era of mining for common disease genes began in earnest.

**

SNPs promised to crack open the mystery of human variation, but taming their power did not come easily. It was to take geneticists another five years to bring them under control. The rationale was simple enough. SNPs were spelling changes in the code. These changes led to diseases. So it was a matter of finding which spelling changes corresponded to particular diseases. The way to do it was to corral about a thousand people affected by a disease like diabetes and a thousand people very similar in age, sex and ethnic background who weren't affected. Then it was just a matter of checking the spelling differences between the two groups. Hopefully (as Cantor thought), a few spelling differences would stand out like a sore thumb. And presto, you would be led to the genes for diabetes.

It didn't work out that way.

For one thing, nobody could afford to do the complete experiment. There are some ten million SNP variants in a human genome, each of which cost at least 20 cents to read.[41] And researchers didn't have the full catalogue of SNPs anyway.

So they took a more a pragmatic approach. Not all SNPs were equal. Some spelling mistakes were common, appearing at that particular address in the code in more than 1% of the population. Researchers led by Eric Lander decided that it would be this more limited subset they would test. After all, they were hunting for the genes behind *common* diseases; it was reasonable to expect that the *common* SNPs would be responsible.[42]

Then a fortuitous discovery meant that geneticists could reduce the number of SNPs they needed to test even further. The previous generation of gene hunters managed to get by with very few landmarks to track genes across DNA because they were tracking genes in a family. Genes in one family are rather like a pack of minimally shuffled cards where the jack is next to a king, queen and ten of hearts. Just as I can be confident that if I locate a jack of hearts in a poorly shuffled pack then the ten of hearts will be nearby; likewise, in families you can be confident that a particular marker will stay linked to the disease gene in all the family members. But out in the general population, things are different. Chromosomes are shuffled— so shuffled that scientists thought a marker linked to a gene in one person would not stay linked to that gene in another person.

But it turns out that the shuffling is not even.[43] Some parts of the genome are well shuffled; others look like they have hardly been shuffled at all. In these unshuffled regions, large numbers of SNPs always travel together in what's called a haplotype block. It took the genome researchers another few years to identify all these haplotype blocks and produce the so-called HapMap: a map that shows which blocks of the genome travel together through populations and which regions do not. (See the box on pages 104–5 for the unexpected racial revelations of the HapMap.) It was worth the four-year wait.[44] In those parts of the genome where the haplotype blocks are small, they still needed as many common SNPs as they could find. But in the large blocks, they could get away with using just a handful of SNPs; one SNP would mark all the rest just as the jack of hearts in my imaginary pack marked a nearby set of hearts.

There was still one more thing needed to complete the taming of the SNPs. Geneticists had to get the statistics right. Preliminary attempts at using the SNPs had caused trouble. One group of geneticists would claim that a particular SNP was associated with a particular disease, musical ability, memory or depression. But no-one else could replicate that finding.[45] The geneticists realised they were becoming ensnared in 'statistical noise'. They were testing tens of thousands of SNPs for their association with a disease. Chance, as the statisticians explained, would ensure that some of them appeared to be linked with the disease. So, the statisticians set a daunting

A set of SNPs

SNPs travel together in haplotype blocks like the cards of an unshuffled pack.

hurdle. No SNPs could be deemed significantly associated with a disease unless that association had less than a one in a hundred million probability of occurring by chance.[46]

Some famous gene associations have bitten the dust this way, most recently an association between depression and a SNP near the gene for the serotonin receptor. This association made a lot of sense. Serotonin is a brain chemical associated with mood: low levels are linked to depression. It works by transmitting messages through the serotonin receptor. So a SNP that affected the receptor might also affect depression. Unfortunately the statistical facts got in the way of the story.[47]

By 2006, the gene hunters had amassed 500,000 SNPs of the common variety,[48] each strategically chosen to mark out the entire landscape of the genome. They also had their statistics worked out.

They were tooled up and ready to ask the question: Did people who carried the same disease share some of the same SNPs?

The answer was a resounding *yes*. At the close of each year, *Science* magazine features a 'breakthrough of the year'. In 2007, that breakthrough was hailed to be 'human genetic variation'.[49] After years of false leads, SNPs were finally proving their worth. In 2007 alone,

Signatures of race

With the first reading of the human genome, the concept of a biological racial signature appeared to evaporate. As Natalie Angier wrote in *The New York Times*:

> Dr. Venter and scientists at the National Institutes of Health recently announced that they had put together a draft of the entire sequence of the human genome, and the researchers had unanimously declared, there is only one race—the human race ... The citizens of any given village in the world, whether in Scotland or Tanzania, hold 90% of the genetic variability that humanity has to offer.[50]

The recognisable differences between races were truly only skin-deep. Pigmentation, for instance, was controlled by only a tiny percentage of our genome. The vast bulk of our genome defied any racial typing. 'If you ask what percentage of your genes is reflected in your external appearance, the basis by which we talk about race, the answer seems to be in the range of .01 percent,' said Dr. Harold P. Freeman, the director of surgery at North General Hospital in Manhattan. Craig Venter, then head of the Celera Genomics Corporation, summed it up. 'Race is a social concept, not a scientific one. We all evolved in the last 100,000 years from the same small number of tribes that migrated out of Africa and colonized the world.'[51]

Between the lines you can detect the almost audible sigh of relief. The politically super-sensitive spokespersons of the genome project wanted to keep a barge pole's distance between themselves and the eugenics movements of the past—from Hitler's genocide to the 60,000 forced sterilisations of the 'feeble-minded' that took place in 30 states of the US from the 1920s.[52]

But the HapMap made the geneticists whistle a different tune: genetic signatures of different populations were detectable. The developers of the HapMap picked four different population groups for their study: Yorubans from Ibadan, Nigeria; northern and western Europeans; Japanese from Tokyo; and Han Chinese from Beijing. And they found that the composition of the genome landscape, so-called haplotype blocks, differed in size between the populations. Among the Yorubans, the haplotype blocks were small. The Europeans and Asians had larger haplotype blocks but a different pattern from each other. The findings made perfect sense. Africa is the cradle of humanity, the place that has seen the most generations of mankind, and so the genomes of the direct descendants show plenty of chop and change. Europeans and Asians are a recent offshoot of the family tree that was spawned by a relatively small group (some estimates say as few as one thousand people, around

150,000 years ago). Their genomes have not yet experienced as many generations and therefore show less chop and change. In the immediate aftermath of the HapMap, geneticists braced themselves for the onslaught that predictably came. Racist groups found sophisticated new ammunition for their hate websites.[53] On the other hand, the ability to detect racial signatures in genomes has led to a new recreational pastime as people trace their ancestry from the mosaic of haplotype blocks visible in their genome scans: see <23andme.com/ancestry/origins/>. But be warned: a friend of mine got quite a shock. He'd always believed he was a pure-bred Italian. After posting his DNA ancestry results on the 23 and Me website, within minutes he got a call from a friend. 'Hey dude; I didn't know you were a Jew.' It turned out one of his haplotype blocks carried a signature unique to European Jews. My friend started probing the skeletons in the family cupboard. The secret finally emerged: his paternal grandfather had been adopted and he was indeed a Jew!

12 common diseases were linked to common SNPs. Either the tiny spelling error was itself the cause of the problem, or the problem lay in a stretch of DNA close by. The star performance was a study from the Wellcome Trust. They analysed 500,000 common SNPs in 17,000 people from across the UK, and found that some of the spelling errors were indeed linked to rheumatoid arthritis, hypertension, manic depression, coronary artery disease, type 1 diabetes, type 2 diabetes and Crohn's disease.[54] Other studies found SNPs that were linked to macular degeneration, breast cancer, restless leg syndrome, atrial fibrillation, glaucoma, amyotrophic lateral sclerosis, rheumatoid arthritis, colorectal cancer, ankylosing spondylitis, autoimmune diseases and even the onset of AIDS (which we will hear more about in the next chapter).

2007 was also the year researchers discovered that genomes vary by a lot more than SNPs. Single-letter changes were just the tip of the iceberg. Chunks of DNA up to a million letters long could vanish, appear in multiple copies, or reverse their direction. These missing or extra chunks are often referred to as copy number variations. Geneticists had only ever seen massive code scrambling like this in patients with serious disorders. But it seems the genome can tolerate quite a bit of scrambling and still produce a 'normal' human being. Previous estimates of the similarity of human beings, based on SNPs, had decided that we were 99.9% the same. After finding how common these chunky errors were, the estimate was revised down: we were only '99.5% the same'.[55]

The SNPs and copy number variations landed the geneticists at an address in the genome. With their genome map of the local area, they could take a look to see what other genes lay in the vicinity. Some of them proved to be quite a revelation. For instance, one copy number variation that showed up in the Japanese population turned out to be multiple copies of a gene for digesting starch—a distinct advantage for people who live on rice.[56]

The revelations were dazzling in Crohn's disease too—a disease where the immune system attacks the gut. One of the SNPs ended up in the vicinity of a gene called NOD2.[57] People who carried this SNP were up to 40 times more likely to develop Crohn's disease.[58] NOD2 doesn't mean much to most of us but it rang bells for the immunologists. It is the alarm system for cells called macrophages— the most primitive contingent of the body's defence force. Their major strategy is to *eat* the enemy. Immunologists realised they better take a closer look at primitive macrophages if they wanted to understand Crohn's disease.

Why would a faulty alarm for macrophages tip people over to Crohn's disease? One theory is that when the alarm is ringing loudly, the macrophage army grows a little deaf to it. In immunological jargon, it develops 'tolerance'. And that can be a good thing. The more tolerant an army is, the less likely it will respond to a false alarm and risk firing at its own gut tissue. A faint alarm, on the other hand, may leave the macrophages straining to hear it, creating an army of jittery, trigger-happy soldiers who do tend to fire at their own tissue.[59]

Perhaps that explains why Crohn's is a disease of the affluent. People growing up in rural India rarely suffer from the disease, probably because their intestines are laden with parasites that keep their alarm bell ringing. In affluent countries, not only are the intestinal parasites fewer, they are also of different types. While the rural Indian intestine is rich in hookworms, the affluent intestine is enriched with bacteria that thrive with refrigeration, like different species of *Yersinia* and *Listeria*. Some researchers believe that particular denizens of the gut do a better job than others of toning down the NOD2 alarm system. While *Yersinia* and *Listeria* are supposed to do a poor job,[60] hookworms are supposed to be particularly good. Some researchers are even using hookworms as a therapy for Crohn's disease![61] The newly discovered NOD2 gene has justified a rather unconventional approach to treatment.

The most spectacularly successful gene hunt was for genes associated with age-related macular degeneration (AMD). If you're past the age of 60 and start noticing a fuzzy black hole at the centre of your gaze, it's probably AMD: a quarter of people over the age of 65 succumb. The disease wipes out the most sensitive part of the retina, a patch of cells called the macula. When the researchers compared the SNPs of people with and without AMD, they found those with the disease were more likely to show a misspelling near a gene called factor H. Factor H takes us right back to the war games of the immune system. Its job is to defuse a piece of ordnance called 'complement'. Complement is a bomb made of 'complement proteins'. They circulate harmlessly in the blood and usually assemble on the surface of bacteria to blast holes in them. But sometimes they attack the wrong target, and then it is up to factor H to defuse the bomb. AMD researchers had long held suspicions about the shady nature of complement proteins because they were found lurking in drusen, a pale yellow substance that coats the retina in the early stages of the disease.[62] Now, finding that susceptible people have misspellings in factor H confirmed those suspicions. AMD was, yet again, a case of the immune system's friendly fire—this time from complement bombs gone awry.

Overall, in people with AMD, the gene hunters found SNP misspellings in three different genes—all of them played a role in regulating the complement bomb. For people carrying the worst SNP varieties of these three genes, their risk of developing AMD was a staggering 250 times greater than those who carried the more protective varieties. This was a rare case where gene hunts identified 75–80% of the genetic cause of the disease.[63]

Crohn's disease and macular degeneration are poster children for the success of the new gene-hunting techniques. The newly discovered genes went a long way to explaining who will get these diseases, and why. But even diseases that had become graveyards for burnt-out gene hunters began to yield their secrets.

Take type 2 diabetes, for instance. The blame for this disease lies squarely at the feet of 'beta cells' that reside on little islands in the pancreas. Their principal job is to keep tight control of the blood's sugar level. After a meal, sugar floods into the bloodstream. The beta cells sense the rise and release the hormone insulin. Insulin acts like a key, activating a conveyer belt for sugar on the outer membrane of cells. As the cells draw in the sugar, the blood level gradually falls. But if there isn't enough insulin released, the conveyer belt fails. Not only does that starve the body tissues, the blood sugar levels rise so high that they poison the blood vessels. This is exactly what happens in type 2 diabetes; for some reason the beta cells fail to release enough insulin. Geneticists knew that inheriting a bad set of genes went 40–50% of the way to explaining who would get the disease. And they had tried to find these genes. But diabetes had earned itself the epithet of 'geneticist's nightmare' for all the false leads thrown up. Rare forms of the disease had provided a few susceptibility genes[64] but when it came to garden variety diabetes, nothing convincing had been discovered. Now the new studies were delivering.

One of the genes that turned up was the FTO gene. It's a gene that also strongly predisposes people to being obese. If you have the wrong two copies of FTO, you are likely to be 2–3 kg heavier than someone with the alternate forms of the gene.[65] Since being obese is one of the strongest risk factor for diabetes, it was no surprise to find this gene.

But some genes were 'not on anyone's radar screen', Mark McCarthy, the Oxford University geneticist who led these studies, told me. One was the gene for the melatonin receptor that forms part of the mechanism of the body clock. Melatonin, a hormone released from the brain's pineal gland each night, works together with receptors on body cells to keep the body ticking away according to the circadian (day–night) rhythm. All very well, but what does the body clock have to do with diabetes? 'Lots,' says McCarthy. As darkness falls, melatonin seeps into our bloodstream and among the many things it does to the body each night, it curtails insulin secretion—which makes perfect sense. Most animals do not eat at night. Perhaps having an overactive melatonin receptor also curtails the regular release of insulin, predisposing that person to diabetes?

Another set of genes appeared to control the ability of the pancreatic beta cells to multiply. Again, these genes were easy to plug into the picture. Pancreatic cells that multiply their numbers might be able to keep churning out enough insulin for a person's needs, no matter how many sweets they imbibe. On the other hand, cells that divide with gusto are a risk for cancer! Indeed, the studies detected a yin–yang phenomenon: some of the same gene variants that were protective for diabetes raised the risk of prostate cancer.

Another surprise was a link between diabetes and a gene that imports zinc ions into the cell, called SLC30A8. Again, there was a plausible explanation. Beta cells carefully protect their insulin stores by binding the molecules together with zinc. People carrying the SLC30A8 variant might have less zinc and their insulin stores could fall apart. 'Genetic studies of this kind are revealing new and unsuspected connections between diseases,' enthused Eleftheria Zeggini, McCarthy's colleague at the University of Oxford.[66]

Schizophrenia, another disease that stymied many a gene hunter's career, also began showing new signs of life. Schizophrenia is about

80% heritable, meaning that the disease can mostly be blamed on the genes a person inherits. But again, researchers had largely failed to find these genes. A handful had been identified in studies of affected families or in isolated populations like Icelanders.[67] One such gene was neuregulin, which governs the way brain cells are wired together. It offered a tantalising explanation of how bad genes might predispose a person to schizophrenia: a faulty neuregulin gene might cause faulty brain wiring, and that could lead to a disconnect between perception and reality.[68]

But did genes like neuregulin play a role in common schizophrenia that affects about one in 200 people in most populations?[69] Researchers got excited when they tested an American population. Comparing the genomes of schizophrenia sufferers with those of unaffected people, they found the schizophrenia sufferers were three times more likely to carry copy number variations, that is, extra chunks of DNA or missing chunks. These chunky errors seemed to concentrate around genes that wired to neuregulin.[70]

The fact that in most cases researchers still only had a dim idea of how to connect the dots between the newly identified genes and how they contributed to diseases didn't seem to daunt them. Worse, in most cases the common SNPs that associated with diseases could not be pinned to any gene at all. Most of these misspellings lay in DNA that seemed bereft of genes—so-called non-coding DNA.[71] For instance, a study of people with multiple sclerosis yielded a weak link to SNPs lying in non-coding DNA near a gene involved in vitamin D metabolism.[72] But that was enough to excite Simon Foote, director of Tasmania's Menzies Institute: 'I've been studying multiple sclerosis for 15 years. Apart from futzing around with HLA genes[73] there's been nothing. Now we have something statistically believable. It *might* give us some interesting biological understanding'.[74]

The mood at the end of 2007 was euphoric. Researchers had started digging out the first nuggets of what promised to be a rich gold mine.

'Are we on the threshold of something? I think so,' enthused Neil Risch, a gene hunter at the University of California, San Francisco.[75]

***

What a difference a year makes! The buzzword of 2007 was GWAS— short for Genome-Wide Association Scan, the tool that had blasted open the gold mine of human variation. But by 2008, a new term had entered unwelcome into the gene miner's lexicon: dark matter. In physics, dark matter connotes a mystery. Rather than flying apart our universe hangs together—presumably because of the attractive force of matter. Yet when physicists measure all the matter they can see, they find there is nowhere near enough to keep the universe from coming unstuck. The undetectable stuff that must be there is dubbed 'dark matter'.

Though the gene mining operations of 2007 had started to deliver, except for the notable exception of macular degeneration, the actual size of what they were delivering was pitiful. Take diabetes, for example. Genes are supposed to explain at least 40% of the disease. But when the gene miners added up all the gene associations they found, they explained just 4%. To do better, the gene miners pooled their data. Researchers from the UK, Finland, Sweden and the US pooled about 10,000 people of European descent. Then they replicated the study in a further 57,000 Europeans.

And the result? Six more gene associations—but each had only a tiny effect. Nineteen genes in total were returned and depending on how you torture the statistics, they account for 5–10% of the genes that predispose people to diabetes.[76] That means that 90–95% of the genes that predispose people to the disease are still missing. It is these missing genes that have been dubbed 'dark matter'.[77]

The story was repeated over and over. Height is a famous example. We know that genes are 90% responsible for determining how height varies in a population, but the genes captured in the miner's sluice explained just 3%.[78] When it comes to coronary artery disease, again several sifted genes accounted for only 2.4% of what researchers expected to find.[79]

In mid-2009, the schizophrenia story got worse. The 2008 studies had pointed the finger at disruptions in brain-wiring genes like neuregulin. But when researchers did the definitive GWAS study, sifting through the genomes of 27,000 people to see what showed up, they found no nuggets. No neuregulin, no genes at all that had anything to do with the brain. The strongest association they could find was for HLA genes of the immune system.[80] Some researchers tried to make the best of it. Limply, they offered that schizophrenia may also be an autoimmune disease and that this explained why people with schizophrenia are slightly more likely to have been born in winter.[81] (Presumably the logic here is that if you are born in winter you are more likely to be exposed to viruses and if you have a weird immune system, it might overreact and damage your brain.) It's not hard to see why these publications created somewhat of an uproar.[82] *The New York Times* labelled it 'a Pearl Harbor of schizophrenia research'.[83]

I first heard of dark matter in February 2009 at a genome conference in Lorne, Victoria, a gorgeous coastal locale that annually attracts flocks of migrating geneticists from the northern hemisphere. I immediately thought of Joseph Terwilliger and another coastal conference about eight years ago. Was Terwilliger now being proved right after all? When I got in touch with him after this eight-year hiatus he had no compunction in saying, 'I told you so.'

> I guess nowadays it is getting pretty hard to find anyone who still thinks GWAS was a productive venture now that the empirical data fits almost too perfectly with what Ken Weiss and I had been saying a decade ago …

And so we're back to the brouhaha of 2008 and its startling quotes. To reprise Duke University geneticist David Goldstein in *The New*

York Times: 'It's an astounding thing that we have cracked open the human genome and can look at the entire complement of common genetic variants, and what do we find? Almost nothing. That is absolutely beyond belief.'[84]

The bewildered gene miners had, in many cases, returned from their prospecting with slim pickings.[85] Their expected booty was locked away in something called dark matter.

***

Dark matter forced geneticists to squarely address a hundred-year-old question: How are common traits put together?[86] Or in more elegant terms: What is the 'architecture of complex traits'?[87] The genes behind these traits were largely invisible to the genome-wide scans. Why?

There were two explanations.

One was that these genes had tiny effects. Like gold dust, hundreds or thousands of them came together to form the trait. The reason so few of the genes had been picked up by the sifters was that they were too easily lost in the background noise. Peter Visscher, a statistical geneticist at Queensland Institute of Medical Research,

Architecture of common traits: gold dust or nuggets?

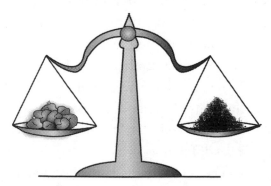

Are the genes behind common traits like gold dust or like nuggets?

is firmly in this camp. He recently went on a hunt for the genes that explain height differences in a British population. Using a novel statistical approach, he calculated that 50% of the genetic predisposition to height lay among genes whose effects were so tiny they barely reached statistical significance.[88] If the gene miners scaled up their operation, he predicted, more of these tiny signals would emerge out of the statistical noise. To some extent, Visscher's prediction is being borne out. A gargantuan genome-wide association study for height genes scanned the genomes of 180,000 people and found another 180 genes. The geneticists are now able to account for 10–13.5% of the heritability of height, up from a previous 3%. This study has one of the longest list of affiliations I've ever seen. Two hundred and three departments and about three hundred authors![89] For Visscher, dark matter has been illuminated—most of it is tiny gold dust lying amid the alluvial sand that just requires more and more sluicing to retrieve it.[90]

That's not the way David Goldstein sees it. In his view, human traits are not built of genetic gold dust but of nuggets. The reason we haven't been able to see them is because we've been using the wrong equipment. We've peered into the dark mine of the hidden genome using a beam that only lights up the common mutations—the common SNPs. But what if human traits are built from rare mutations? The beam won't illuminate them; they will in fact be 'dark matter'. In other words, you and I may both be highly predisposed to diabetes. But my mutations may be completely different from yours. If we're both placed in a study looking for mutations in common, nothing much will emerge. Goldstein backs up his view that disease mutations are rare with a compelling example. Epilepsy can be caused by a defect in a single gene and when that happens multiple members of a family are affected. The University of Melbourne's Sam Berkovic, the doyen of epilepsy genetics, has studied such families for 20 years. In nearly every case, he's found a different gene is responsible. Says Goldstein, 'The only thing we can be sure about is heterogeneity; different genes can contribute to the same disease.'[91] Geneticist Mary-Claire King at the University of Washington, Seattle—famous for the discovery of the rare hereditary breast cancer genes BRCA1 and BRCA2—is in Goldstein's camp. For her, Tolstoy's description of unhappy families is an apt way to describe the genetic make-up of human disease: 'Every unhappy family is unhappy in its own way.'[92]

To sum up, the Visscher camp thinks human traits are built of common genetic dust; the Goldstein camp, of rare nuggets. Right now we can't tell which is right. 'We did all these genome-wide association scans but in a way they didn't tell us much. We have not resolved the architecture,' Goldstein told me.[93]

To find out, we need to go down into the dark mine of the human genome and start reading every letter of the six billion–letter genome—which is exactly what the gene miners are doing, and at breakneck speed.

Goldstein told me he had already fully sequenced the genomes of 151 people with schizophrenia, 100 with epilepsy and a handful who are resistant to infection by HIV. Not to be outdone, Mark McCarthy told me that he had already fully sequenced the genomes of 750 people with type 2 diabetes and another 750 without it. It's happening everywhere. 'We're seeing a flood,' Goldstein told me. McCarthy agreed. 'The world's in a frenzy. In 3 to 5 years we will have read genomes from nose to tail; we'll know all the common and rare variations, all the copy number variations. We'll have a pretty good idea of what they will explain.'

As early as 2014, the flood of genomes might at last tell us how genetic traits are built. And then will we be at *Gattaca*?

***

It depends on what we find. For diseases at the nuggetty end, where a handful of genes are involved, prediction is going to be a real possibility. For those at the dusty end, researchers differ in their prognoses. Visscher thinks it should still be possible to weigh up how much genetic 'dust' a person inherits to arrive at a good prediction of their risk. Indeed, he cites studies in animal breeding that have successfully used this approach. The best dairy cows were successfully predicted based on the total amount of genetic dust they inherited.[94]

Others think it won't be so easy because genes may interact with each other in strange ways: it won't just be a matter of adding up what's there. For instance, you see this in fruit flies. There is a bizarre mutation that puts a leg on a fly's head; it's called antennapedia, or, as they say in the lab, 'nice legs shame about the face'. But geneticist Greg Gibson, now at Georgia Tech University in Atlanta, found that the mutation disappeared when he crossed his inbred flies to another strain. Even though these flies carried the dominant gene, the different mix of background genes completely cancelled the effect of the mutation. The fruit fly shows that any one gene can have totally unpredictable effects depending on the company it keeps. 'In my view it's not a *Gattaca* situation; I don't think we'll ever get to prediction. If I gave you the genomes of Craig Venter and Madonna you would not be able to pick who was brighter and who was more musical. I find that an empowering thought,' Gibson told me.[95] Researchers have found similar things in mice—genes do different things depending on who they're partnered with.[96]

And human beings? The jury's out. There are certainly examples of the same gene doing different things in different people. In a family that carries the hereditary breast cancer gene BRCA1, mothers, daughters, sisters and aunts who carry the gene are at very high risk of developing the disease. But these family members don't just share BRCA1, they also share many other genes that bring out the worst in the inherited cancer gene. In the general population, researchers have found the same BRCA1 gene can be carried by women who do not have such a high risk of breast cancer. Here the gene seems to be in better company; these women carry genes that mitigate the bad influence of BRCA1.[97]

If gene interactions are so unpredictable, how can any genetic test predict disease? If there are indeed a thousand different genes that contribute to schizophrenia and they interact in different ways, then, as McCarthy says, 'It will be a conundrum, we don't have enough people on the planet to do the experiment.'[98] So even if we read every letter of a genome, we may not end up with useful predictive tests.

Nevertheless, the relentless march of technology does seem to be driving us towards the *Gattaca* scenario. We are heading for the time of complete genetic readouts at birth. There's even a $10 million X PRIZE—the prize used to entice futuristic technologies like the first re-usable space shuttle—waiting to reward the first company that can do it quickly, economically and en masse.[99] 'The future that we all envisage is the day when every infant has their genome sequenced at birth and we utilise that information to optimise health throughout their life,' beamed Andrew Wooten from the X PRIZE Foundation in a television interview.[100] Predictions of the $100 full genome are rife: that's about the current cost of the gene test offered by '23 and Me' which only scans about a million letters of your genome.[101] Some say that, like mobile phones or gmail, genome sequencing will eventually be available for free, part of a business model that will sell you services based on your genome, perhaps drugs or diets tailored to your health needs.[102]

But as to whether we can decrypt what those genomes mean, we will have to wait a little longer to find out.

Barack Obama's adviser Eric Lander, for one, seems pretty confident that the secrets of the genome will be revealed. 'There are a lot of smart young scientists—if we give them the data they'll figure it out. I doubt it's so complex they can't understand it. Geneticists of the 1950s would have been aghast at the thought of dealing with a million genetic variations: now it's routine. The next generation comes along and takes for granted what the previous generation thought was too complicated.'[103]

5

Surviving AIDS

In 1981, Acquired Immune Deficiency Syndrome descended like a biblical plague upon the gay communities of New York, San Francisco and Los Angeles. Young men in their prime were struck down with a mysterious and terrifying collection of symptoms. There were the stigmata—purple-brown skin blotches that turned out to be Kaposi's sarcoma, a rare cancer usually only seen in old men. They were also ridden with fungal sores and pneumonia caused by garden variety microbes. All were symptoms of an immune system in collapse. The young men died within months.

In 1983 researchers traced the cause of the syndrome, which soon became known as AIDS, to a virus—Human Immunodeficiency Virus or HIV.[1] Yet for 15 long years there was no effective treatment and no vaccine. Millions died, their doctors looking on helplessly. Finally, in 1996, a three-drug cocktail turned the tide. AIDS was no longer a death sentence.

Fast-forward to 2011 and the 30th anniversary of the AIDS epidemic. If you live somewhere like Australia, HIV and AIDS have largely fallen off the radar screen. The Grim Reaper television campaigns of the 1980s are gone. Less than 0.15% of the population is infected,[2] and those living with AIDS have access to ever-improving drugs that promise something close to a normal life.[3]

Things are very different in sub-Saharan Africa, the countries that lie below the belt of the Saharan desert. This is the epicentre of the AIDS epidemic; two-thirds of the world's infected population is here.[4] And in some countries the infection rates are staggering.[5] Take Botswana: a quarter of the adult population has HIV.[6] Yet of all African countries, Botswana stands out as one of the most progressive. It boasts a stable democratic government, national wealth from its diamond mines and, until HIV struck in 1990, its citizens had a life expectancy of 65 years, the highest of any African country. In 2002, it was also the first country in the world to roll out free HIV drugs. Within six years, close to 90% of those diagnosed with AIDS were receiving medication.[7]

So, when I visited Botswana in February of 2008, I imagined that its hospitals, like their Australian counterparts, would be largely free of AIDS patients. I was wrong. I spent a day at the Princess Marina Hospital in Botswana's capital, Gaborone, and was confronted with wards full of dying people. At least in the men's ward they were contained in beds. The women's ward was so crowded they were also lying on the floor.

It was a window into the realities of HIV in sub-Saharan Africa: 22.5 million people are infected and the majority, 60%, are women. No-one really knows why the African epidemic has been 'feminised'. Teenage girls bear the worst burden: they are two to three times more likely to be infected than teenage boys.

That day in the ward made me see that drugs are not enough. With such a large chunk of the adult population infected, and most people unaware of the virus they harbour, the epidemic has built up an unstoppable head of steam.[8]

That is why researchers on the frontline are battling to develop a vaccine, a cost-effective one-off jab that would stop the virus in its tracks, just as a vaccine for smallpox did. They are also trying to find an answer to a question. Why did the AIDS epidemic pick the countries of sub-Saharan Africa for its worst excesses? There are plenty of other places in the world that are poor (certainly much poorer than Botswana), and plenty of other places with a promiscuous sexual culture. Yet nowhere else in the world do you find HIV infection rates surpassing 4% of the population.

In epidemics of the past, researchers looked for answers by training their microscopes on the invading microbe. But HIV foisted itself upon humanity around the same time we starting reading our genomes. This time around, the microscope has also been turned on ourselves.

**

Why is AIDS so different in sub-Saharan Africa? Max Essex from Harvard University School of Public Health, has spent over 15 years trying to find out—some of them peering into the genomes of the people of Botswana. The greatest human genetic diversity on earth is found among Africans. Perhaps those genetic differences explain why some of them are so susceptible to HIV? If Essex can find the genes responsible, it will give researchers new leads to fight the epidemic.

Bruce Walker, Director of Harvard's Ragon Institute, is also peering into genomes for answers. He is part of a global effort trying to develop an HIV vaccine. But so far, their efforts have been a dismal failure. Three different vaccines developed over 20 years have failed to teach the immune system to defeat the virus. Yet about one in 200 people, so-called 'elite controllers', do it without any training. Walker is probing their genomes to find out how, and to use that information to make a better vaccine.

Essex and Walker were not the first to believe that the key to winning against this virus lies within our genes. That honour goes to Stephen O'Brien, Chief of the Laboratory of Genomic Diversity at the US National Institutes of Health.[9]

**

If you were parachuted into O'Brien's lab at the National Cancer Institute in Bethesda, Maryland, you might think you'd landed in a zoo. Strolling along the rows of large white freezers, you'd see they stored hundreds of thousands of tissue samples from Serengeti lions, Chinese pandas and orangutans from the forests of Borneo, not to mention cheetahs, leopards, elephants, mice and … people. Why is an HIV researcher studying such a menagerie of animals?

The short answer is that O'Brien is a gene hunter. Indeed, he's been dubbed 'the Indiana Jones of genomics'. That's partly because at 60-something he is ruggedly good-looking and has an insatiable appetite for discovery. But mostly it's because of his spectacular hunts through the genomes of wild animals. It's estimated that over 99% of all species that ever lived have become extinct.[10] The winning strategies are recorded in the survivors' genes. It is these survival genes that O'Brien is hunting.

Today's human populations are also survivors. Before the era of vaccines and antibiotics, the statistics were bleak. Consider, for example, bubonic plague, carried by fleas and rats. In the Justinian plague of 541 it culled the population of Europe by half; in 1347 it returned for a repeat performance as the Black Death, and again in 1665 as the plague of London. We have also been ravaged by small-pox, malaria, tuberculosis, typhus, influenza, cholera, polio and yellow fever, to name a few. Yet we are still here because certain individuals survived and passed their genes down the generations. So what are some of these survival genes? Until now we've had only the tiniest inkling. Some of the most commonly inherited gene mutations are the sickle cell haemoglobin gene, present in 5% of the world's population (but up to 10–40% of populations across equatorial Africa);[11] the cystic fibrosis gene, present in 5% of Caucasians;[12] and the Factor Five Leiden gene, present in 3.3% of Caucasians.[13] When inherited from both parents, these mutant genes cause devastating diseases—anaemia, cystic fibrosis and deadly blood clots—so why has evolution allowed them to become so common? The reason is that when inherited from only a single parent, these genes have survival value. A single dose of the sickle cell haemoglobin gene protects against malaria (which is why it's so common across malaria strongholds like equatorial Africa); the Factor V Leiden gene protects against blood poisoning and possibly plague;[14] and the cystic fibrosis gene might offer protection against cholera.[15] There must also be less conspicuous genes that gave survivors an edge but didn't exact such a deadly bargain. In the era of genome mining we ought to be able to find these genetic gems. At least that's what Steve O'Brien thought.

The fact that O'Brien started his scientific career as a fruit fly geneticist also explains a lot. They are the ones who pioneered the art of mining genetic treasures from the genome. Their starting point was usually an enticing mutation, a sort of 'X marks the spot'. For instance, rare fruit flies with white eyes rather than red, provided a trail to a gene that controls eye colour. It's rather like an olden-day mechanic searching for the reason why a car wouldn't start. Tracing his way through the engine, he'd eventually hit on a faulty spark plug. As a junior scientist, O'Brien spent several fertile years as a fruit fly gene miner, but in the late 1970s he sniffed a new type of genetic quarry. In animals, cancer is sometimes spread by a virus. Not all animals succumb equally; some appear to carry resistance genes. Over the years, O'Brien helped hunt down these virus-resistant cancer genes. For instance, a population of mice in Lake Casitas, California, successfully fought a viral epidemic for years. Their protective gene turned out to be a pruned down version of the virus itself![16]

When AIDS exploded onto a helpless human population in 1981, O'Brien heard a call to arms. Somewhere in the vast genetic mix that makes up a human population, there would be resistant individuals who would lead him to genes that could fight off the virus. As he wrote in his book, *Tears of the Cheetah*, 'I was never very confident that medical scientists could guess at all possible therapeutic approaches for deadly diseases; I was more optimistic that countless generations of trial, error and natural selection had come up with innovative genetic solutions for deadly disease.'

O'Brien also had a very personal connection to the epidemic. His brother had died of AIDS.

***

In 1984, the gene hunter donned his gear and set off into the jungles of American cities, hunting for rare individuals who had natural resistance to HIV. He had tantalising hints that they existed. For instance, before routine screening of the blood supply began in 1984, 12,000 American haemophiliacs received contaminated blood products. Eighty-five per cent of them became infected, but 15% did not. Over the years, O'Brien amassed ten thousand blood samples from people who had been exposed to HIV, including haemophiliacs, gay men, intravenous drug users and HIV-positive mothers and babies. And from this blood, his colleague Cheryl Winkler extracted the DNA to provide a representation of the genome of each individual.

Just as he had expected, individuals infected with the virus responded differently. Some progressed within months to full-blown AIDS; others remained healthy for years. And when the virus finally decimated people's immune systems, different diseases took hold. Some got Kaposi's sarcoma, others pneumonia or meningitis or lymphoma or dementia or some combination of these. Some lucky people *never* became infected, despite being transfused with infected blood or having regular sex with an infected partner.

O'Brien sorted the study participants into different categories, in much the same way he had sorted flies according to eye colour in a previous life. The variations marked the head of the trail, but how was one to trace it back to the genes responsible? In the 1980s, finding human genes of any kind was no easy business. Only about a thousand had been identified; and the entire set was estimated at around 100,000.[17] If genes for HIV resistance truly existed, how was he to find them? Like any good hunter would: by knowing the habits of his quarry. O'Brien zeroed in on the territory where HIV resistance genes were likely to be found.

***

HIV is a Trojan horse that tricks the cell into allowing it entry and then hijacks the cell's equipment to make more copies of itself. To gain entry it latches onto one of the cell's docking stations (CD4 receptor) using a three-pronged grappling hook (GP120). Once hooked on, the virus injects its payload into the cell: two copies of its RNA-based genetic code and a few enzymes. Now it's single-minded aim is to splice its tiny code of nine genes into the cell's genome. But first it has to rewrite its RNA code into DNA, the format of the host's genome. It does so using its 'reverse transcriptase' enzyme. With the rewrite completed, it slips into the host genome and the battle for the control of the cell is virtually over. Because while the cell goes about its day-to-day business of copying DNA recipes into RNA, it also ends up making millions of copies of virus RNA. Adding insult to injury, the newly copied viruses help themselves to the cell's own proteins to cushion their flimsy RNA genome. In a final act of brazenness, they snatch bits of the cell's membrane to package themselves for export. The doomed cell is now a virus bomb. As it explodes, millions of viruses burst forth ready to invade again.

O'Brien figured the brazen habits of the virus might also be its weakness. It was utterly reliant on the human cell's copying machinery. Just as Canon and HP copiers have slightly different performance features, people also carried different genetic 'brands' of copier. Perhaps carrying a particular brand of copier could give people an edge over HIV?

The Human Genome Project was helping to pinpoint these genetic brand variations. At most places on the DNA code, human beings have exactly the same letter. But minor misspellings, known as SNPs (single nucleotide polymorphisms), occur about once every three hundred letters, rather like spelling the word aluminium as aluminum. Perhaps these misspellings made a difference when it came to resisting a virus? Michael Dean in O'Brien's lab checked to see. He and O'Brien tested hundreds of spelling variations in and around the copier genes to see if any of them were associated with

How HIV hijacks the cell's machinery

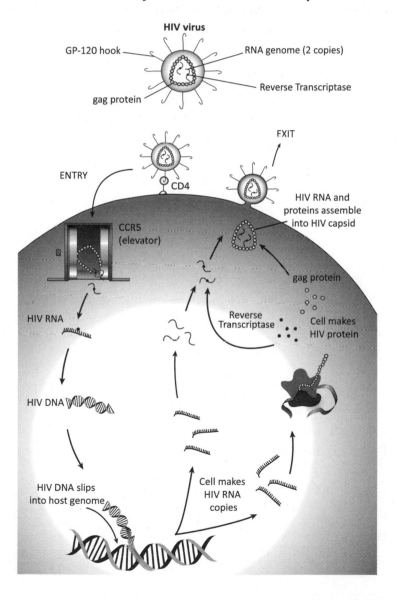

either a better or worse response to HIV. Twelve years and one million dollars later, they had *nothing* to show for their efforts.

Finally, in 1996, they got a break. Apparently just hooking onto a docking station was not enough for HIV to enter into the cell: it also needed an elevator to take it down. That elevator was the CCR5 co-receptor.[18] One of its discoverers, John Moore at the Aaron Diamond Research Institute in New York, described how CCR5 and CD4 work together with a vivid metaphor:[19]

> Imagine HIV as an airborne blimp full of people en route to the subway tunnels below New York City. The Empire State Building represents the CD4 molecule sticking up from the cell, that is, Manhattan. The blimp hooks to the building steeple and waves in the breeze. Eventually a large cable car elevator, the CCR5 molecule, adjoins the dirigible, off-loads the human cargo and delivers them downward to the subway station below. Manhattan Island—the cell—has become infected with HIV.

It was CCR5 that delivered HIV into the cell. CCR5 also provided deliverance to O'Brien and his lab. When they tested to see whether the people in their study showed spelling changes around the CCR5 gene, they hit gold.[20] Some of the people who resisted being infected by HIV carried a striking spelling alteration; the gene was missing a 32-letter chunk![21] Named CCR5 Δ32, this mutation produced a broken elevator. More and more papers bore out the findings: being born with two defective copies of the CCR5 receptor provided almost iron-clad protection against HIV infection,[22] while a single copy delayed the onset of disease. Yet people who had no functioning CCR5 seemed entirely healthy; there seemed to be no price to pay for a broken CCR5 receptor.

Around 10% of Europeans carry a single copy of the defective CCR5 Δ32 gene; 1% carry two copies, reflecting the odds of inheriting the gene from both parents. The rate of inheritance plummets as you go south. The highest incidence is found in Scandinavia, Finland

and Northern Russia, where it reaches 16%. France, England and Germany hover at 10%; Italy, Turkey and Bulgaria at 5%. By the time you reach Saudi Arabia and sub-Saharan Africa, the incidence is zero.[23] This southern decline intrigued scientists. Perhaps the CCR5 mutation reached such high levels in northern Europeans because it protected them from some historical plague? It couldn't have been HIV—that virus crossed from chimpanzees into humans sometime in the 1900s, while the CCR5 mutation has been common in northern Europe for hundreds if not thousands of years.[24]

Steve O'Brien suspected that the CCR5 mutation might have protected the populations of northern Europe from the bacterium that causes bubonic plague, *Yersinia pestis*. It was a hunch that took him to the tiny English hamlet of Eyam in Derbyshire, a hundred miles north of London (coincidentally, the setting of Geraldine Brooks' novel about the plague of London, *Year of Wonders*). Eyam's population was halved during the 1665–66 plague of London and some of the modern-day residents are direct descendants of plague survivors. O'Brien checked their genes and discovered that 15% of them carried the CCR5 Δ32 mutation, *slightly* more than the general British population frequency of 10%.[25] But whether the mutation was responsible for their ancestors' surviving the bubonic plague remains a subject of debate.[26]

***

Whether or not the CCR5 mutation conferred resistance to plague, it certainly confers resistance to HIV infection. And unlike the sickle cell anaemia or cystic fibrosis mutations, it seems to have no downside, so drug developers wasted no time trying to copy nature's winning strategy.

In 2007, Pfizer released Selzentry™, an oral drug that blocks the CCR5 elevator. Eleven years after the discovery of CCR5 Δ32 and

23 years after Steve O'Brien began his gene hunt, it was the first trophy of the post-genome era, and remains to date the most stunning one.

But the drug is not a perfect fix. Selzentry™ is a key addition to the therapeutic arsenal, however, as with all HIV drugs, some strains of virus may become resistant to it. This tends to happen when people have been infected for many years; HIV evolves to use different elevators and the drug cannot block all of them.

Nature's experiment showed that a blocked CCR5 receptor is most effective at preventing HIV infection in the first place. So another strategy is to load Selzentry™ into a gel that is smeared on the outside of the penis or vagina before sex. Trials are underway.[27]

After the CCR5 discovery, the doorway to finding more HIV resistance genes flung open. CCR5 itself provided a clue. Though it ferried HIV into the cell, clearly that was not what it was designed for. It turns out the regular clients of the CCR5 elevator are chemical alarm signals called chemokines. When there are plenty of them around, they crowd the elevator and HIV has a harder time getting into the cell. When O'Brien's colleagues looked at the genes of people who progressed more slowly to AIDS, they found spelling changes that increased the output of these elevator-hogging chemokines.[28]

Another set of genes consistently showed up as important in determining how rapidly people would progress towards AIDS. And these were no surprise at all: the HLA genes.

***

A human body contains about a trillion cells that could be considered citizens of a nation, the nation of Elizabeth Finkel, for instance. Every one of my cellular citizens wears an Elizabeth Finkel uniform. It's made from six HLA proteins.[29] And as Wikipedia will tell you, 'a uniform is intended for identification and display', which is exactly what the HLA proteins do.

Cells in uniform

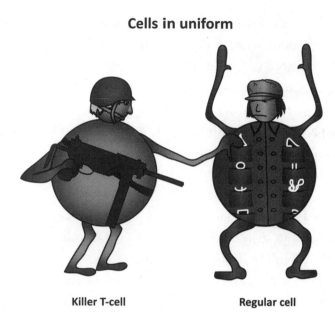

Killer T-cell   Regular cell

Cells wear uniforms made of HLA proteins. The pockets display whatever the cell happens to be making. Killer T-cells frisk the pockets. If they detect a bit of virus, a summary execution takes place

My cells need to wear these uniforms because my outwardly peaceful life is actually being played out in a war zone. The soldiers of my immune system are on constant patrol against the viruses, bacteria and parasites that are trying to invade my cells.[30] Part of their job is to inspect my cells, which in turn must identify themselves as nationals by *displaying* their uniforms to the patrolling forces. The soldiers who specialise in the art of detecting infected citizens are called 'T-cells'. They function as a team: 'killer' T-cells kill infected cells and 'helper' T-cells—the ones infected by HIV—recruit other arms of the immune system. But before the T-cells execute any action whatsoever they must carefully interrogate the cell and they do it by frisking the cell's uniform. HLA uniforms have a way of displaying bits of whatever is being made in the cell. The highly trained T-cells know to ignore the detritus of daily life: bits of household proteins elicit no more response than your snotty tissue does as you get frisked at the airport. But if the T-cell finds a piece of virus, it's like the airport frisker finding traces of explosive on your clothes. All hell breaks loose. The infected cell is destroyed on the spot.

The ability of T-cells to detect an enemy depends on how well they see what's displayed on the cell's uniform. It turns out that everyone's uniform is slightly different,[31] especially when it comes to how well they display bits of virus. So it was no surprise that O'Brien found a link between a person's HLA type and the rate of progression to AIDS.[32]

But HIV is a very devious virus. In its attempt to avoid being displayed, it tries to disrobe the cell of its HLA uniform altogether.[33] That can foil T-cells but not another contingent of the immune army. 'Natural killer' cells detect these cells-out-of-uniform and destroy them. They rely on frisking cells using a sensitive probe.[34] People who progress more slowly to AIDS often carry an exceptionally finetuned probe on their natural killer cells.[35]

Over the last 12 years or so O'Brien and other researchers have unearthed 20 genes that influence how a person will respond to HIV. Some of them reveal an arms race between the genes of virus and host that was undreamed of. O'Brien refers to it as 'ratcheting evolution'. It reminds me of the contests between the cartoon characters Coyote and Road Runner.

***

Most parts of the viral machine revealed their workings within the first few years but there was one part that eluded the scientists for over 20 years. It was a gene that went by the name of Virion infectivity factor or Vif. Vif was a crucial piece of the virus's equipment. Without it, new copies of the virus came out corrupted. What was going on? It was a script that could easily have been written for Coyote and Road Runner.

Let's imagine the scene depicted in the cartoon below, where Coyote the virus is doing battle with Road Runner the human cell.

Enter Coyote virus, who proceeds to commandeer Road Runner cell and churn out copies of his genome.

Road Runner deploys an editor protein[36] to scramble Coyote's genome.[37, 38]

Coyote whips out a gadget called Vif, which stamps a disposal sign on the editor protein dispatching it to the trashcan.[39]

Road Runner protects its editor by spraying it with non-stick spray so the disposal tag won't stick.

Road Runner editor obliterates Coyote's genome.

Bad luck, Coyote![40]

Now let's move to a different scene. Here the major actor is a gene called Trim5 alpha. HIV's flimsy RNA genome needs protection, so it cushions it in proteins stolen from the host.[41] Trim5 alpha trims the cushions back off, leaving the virus genome naked and vulnerable. It's a great defence against HIV—if you happen to be a rhesus monkey.[42] Their nifty Trim5 alpha gene gives them foolproof protection. But for most people, Trim5 alpha seems to have two left thumbs.[43] However, some people who progress more slowly to AIDS carry a different spelling in their Trim5 alpha gene. Perhaps their gene is a little more adroit at trimming the cushions off HIV?[44]

To date, some 20 human gene variants have been found that modify the course of HIV infection.[45] Notwithstanding their enthralling antics, they only go a small part of the way to explaining why some individuals keep HIV at bay. In fact, O'Brien can put a figure on it: they explain about 10%.

For O'Brien, who has now spent over 26 years on the trail, that is an 'unpleasant reality'. Some of the response to HIV may not rely on the host's genes at all but on those of the virus. For instance, in Sydney, Australia, several HIV-positive people infected by a blood transfusion remained healthy for over a decade. They turned out to carry

Gene Wars

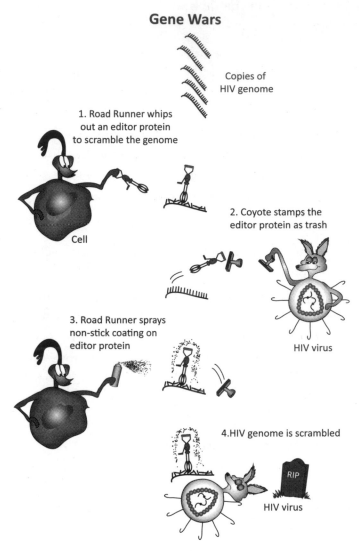

Copies of
HIV genome

1. Road Runner whips
out an editor protein
to scramble the genome

Cell

2. Coyote stamps the
editor protein as trash

3. Road Runner sprays
non-stick coating on
editor protein

HIV virus

4. HIV genome is scrambled

RIP

HIV virus

Like the battles between coyote and road runner, the genes of HIV and the human cell battle for supremacy. The human cell (roadrunner) scrambles the HIV genome using its editor enzyme (apobec). The virus (coyote) fights back with a gadget (vif) that stamps a trash sign on the editor enzyme. Roadrunner fights back by spraying non-stick spray on the editor enzyme. Coyote virus ends up with a scrambled genome and goes belly-up.

a defective form of the virus.[46] And of course the human genome is hardly the only player—factors like nutrition, environment, life history and exposure to other diseases may affect a person's ability to fight HIV.

Still, O'Brien is far from discounting the value of his genetic gems. 'It's true I'd rather have discovered genes that explain 60% than 10%, but 10% is important.'[47] Undaunted he has returned to the hunt, this time using the latest in the arsenal of gene hunter's equipment. For 20 years, O'Brien and his colleagues laid skilful traps by making educated guesses about what the genes were like. Modern gene hunting is more like a mining operation. The researchers make no assumptions about what they are looking for; they simply sift through the mountains of ore to find the genetic gems. And on this expedition, the once-lonesome gene hunter has a lot of company. The latest group to join the hunt are the vaccine researchers.

**

HIV seems to have learnt its tactics from the handbook of urban guerrilla warfare. Scenes from the movie *The Hurt Locker* come to mind.[48] The movie plunges us into Iraq for a close-up view of the operations of an American ordnance disposal team. We watch the heavily armed GIs scan the street warily; they have no idea who the enemy is. Is it the man training his movie camera on them, a friendly teenager, a procession of people pushing a cart, the sheik using his mobile phone?

HIV similarly wears many disguises as it mutates its surface proteins. The GIs of the immune system may be chasing men with movie cameras. But that's yesterday's disguise. Today, the virus may be dressed like the friendly teenager who shimmies up to a group of unsuspecting troops and detonates the bomb hidden under his shirt. Indeed, HIV makes a specialty of infiltrating and destroying the very cells designed to fight it[49]—all of which explains why our immune system can be totally overwhelmed by the virus.

***

'We hope to have a vaccine ready for testing in about two years ... Yet another terrible disease is about to yield to patience, persistence and outright genius,' pronounced Margaret Heckler in 1984.[50] A virus had just been nailed as the cause of AIDS and the US Secretary of Health and Human Services had every reason to be confident of a quick victory. Vaccines had vanquished the polio virus and rendered the smallpox virus virtually extinct.

Heckler's hubris must have angered the gods. Twenty-six years after her fateful prediction, three AIDS vaccines have been tested in human beings. None can claim victory. The first failed. The second was worse than a failure—it increased the rate of infection! The third offered a smidgeon of protection that all but disappeared if the data were analysed unsympathetically. In 2011, there is no vaccine in sight.

The vaccines that vanquished polio and smallpox worked by training the immune system to produce 'neutralising' antibodies. Like guided missiles these antibodies zero in on blood-borne viruses, destroying them before they get a chance to hide away in cells. But so far, no antibody has been able to 'neutralise' an HIV infection.[51] It doesn't happen naturally and it doesn't happen with vaccination, as a trial in 2003 showed. People were injected with fragments of the virus's three-pronged grappling hook (GP120). The hook fragments showed the immune system what to go after and antibodies zeroed in. But they just couldn't get a grip. In the intact virus, parts of the grappling hook are sheathed by sugar molecules while others are folded away like the landing gear of an airplane—only unfurled at the last moment when the virus docks with the cell.[52] Tiny bits of the grappling hook are accessible but they change their appearance so rapidly that antibodies end up chasing yesterday's target. HIV has the stealth capability of an SR71 spy plane!

The next round of vaccines engaged T-cells that can detect viruses lurking inside cells. Bits of virus end up displayed on the infected cell's HLA uniform. T-cells frisk these uniforms and when they detect a foreign shard of virus, it rouses them to battle. To train T-cells, the vaccine developers presented the immune system with a mock target. It was the common cold virus, adenovirus 5, engrafted with three HIV genes.[53] The hybrid virus infected cells, and the shards of HIV did indeed rouse the T-cell army—but with a shocking and catastrophic result. In 2007, it emerged that those people injected with the vaccine were more easily infected with HIV! No-one is quite sure of the reason why.[54]

The third clinical trial used a combination of both types of vaccine, one to recruit antibodies; another to recruit T-cells. At least those vaccinated were not found to be at greater risk, and there seems to have been evidence of slight protection. But nobody is jumping up and down about the success of this 2009 trial.[55]

After the 2007 disaster, there were calls to give up on vaccine development altogether. Many researchers were extremely pessimistic. Nobel Prize winner Peter Doherty told me that as a consultant for the International Aids Vaccine Initiative, he tried to dissuade them from trying vaccines in humans: 'The more we know about the virus the worse it gets.'[56] Ian Gust, an Australian vaccine developer and past director of Melbourne's Burnet Institute, was chair of the scientific advisory committee for the International AIDS Vaccine Initiative when I interviewed him in 2008. He told me, 'All up there are 25 to 30 candidate vaccines being developed; I have no great hope for any of them.'[57]

But there are also stalwarts. Like terrorism, HIV is here to stay. Many researchers feel there is no option but to continue the quest for a vaccine. Anti-AIDS drugs are a great breakthrough, but providing lifelong drugs for 34 million people, two-thirds of them in sub-Saharan Africa, is a logistical and financial nightmare. Only 36% of people who need drugs are getting them.[58] Doubling that figure is predicted to cost US$35 billion annually, three times the current bill.[59] Even then, more than a million new people will be infected each year.[60]

A one-off jab that could protect people for life is still the Holy Grail. And there is one thing that continues to stoke the belief that a vaccine is possible. About one in two hundred people can mobilise their immune system to fight the virus. Some of these so-called 'elite controllers' have been keeping the virus at bay for 30 years. If their secret can be revealed, maybe it can be copied.

To date, no-one has successfully copied an immune response this way. Conventional vaccines aren't much more sophisticated than the one used by Jenner when he injected people with cowpox pus to protect them from smallpox.[61] The vaccines that vanquished smallpox, polio and measles were just weakened or killed variants of the virus. It's enough to wave these guys under the nose of the immune system and like a pack of hunting dogs, it does the rest. In the case of HIV, a wave under the nose is not enough. The vaccine developers are now aiming to take command of the immune army and to do that they need to gather intelligence about what a successful battle looks like. That's where the elite controllers come in.

***

Perhaps it's the suit. Perhaps it's the reserved demeanour. Add the cropped hair and round glasses and Bruce Walker comes across as more corporate than clinical. It's true that he runs a large operation at the Ragon Institute at Massachusetts General Hospital, and helps run a smaller one at the Nelson Mandela School of Medicine in Durban, South Africa.[62] But Bruce Walker is a doctor whose life's work has been devoted to the battle against HIV. It was a destiny forged in an emergency room at Massachusetts General Hospital 30 years ago.

> My first encounter with AIDS came in 1981. A patient arrived at the emergency room with three simultaneous life-threatening diseases, something none of us had ever seen. As an intern, this was an eerie, memorable moment, the first time the people I considered the smartest in the world—the physicians training us—were stumped too.[63]

Serving on the frontline of an epidemic unknown to medical science galvanised the young intern and launched him into a specialty long out of fashion for American doctors—infectious disease. He also trained as a laboratory scientist. Shuttling between patients in the mornings and test tubes in the afternoons, he began accruing intelligence about this battle between humans and HIV. His 30-year quest has taken him to a lot of dead ends. For instance, in the late 1990s, it looked like a temporary round of HIV drug treatment might allow the immune system to gain the upper hand, with reports that people who took 'treatment holidays' seemed to stay healthy—*at first*.[64]

Walker carried out clinical trials and showed that the virus had not taken a holiday. Instead, it had quietly ramped up its numbers and wiped out more T-cells, hastening the person's demise to AIDS. Treatment holidays are now considered a bad idea.[65]

In the main, Walker's research focused on trying to discover the hallmarks of a useful immune response. In the battle between humans and viruses, different tactics work for different bugs. For instance, when it comes to polio virus, antibodies can stop the virus in its tracks. On the other hand, it takes T-cells to bring the herpes virus under control.[66]

What contingents of the immune system were the most effective at battling HIV? Early on, researchers put their hopes in antibodies. They didn't do the trick. Next they put their hope in T-cells. Starting with a key paper in *Nature* in 1987,[67] Walker became famous for dozens of papers showing that T-cells do indeed battle HIV. In the test tube, killer T-cells isolated from patients' blood did a tidy job of destroying HIV-infected cells. He also showed that the adjutants of this fighting force—the CD4 helper T-cells—were coordinating the battle by rallying other contingents of the army. Maybe it was just a matter of giving the T-cell army a head start? That idea led to the world's first vaccines aimed at mobilising T-cells. As we've seen, that also turned out to be a bad idea. Simply directing T-cells against HIV was not going to work for this virus.

For the first ten years, Walker could only pore over the logistics of failed battle plans. But in 1992 he got his first inklings of what a winning battle might look like.

***

People who carried the CCR5 Δ32 mutation never battled the virus. They blocked the enemy at the gates so to speak, and they were easy to identify. Despite repeated exposure to the virus, they never became infected as evidenced by their negative antibody tests.

Another group of survivors was not so easy to identify from the outset. They did become infected but they battled the virus—so successfully, that unlike most people, they did *not* succumb to AIDS within the first ten years. These rare outliers on the bell curve were called 'non-progressors'.

Walker first heard about them in 1992, when he got a call from Susan Buchbinder, an epidemiologist at the University of California, San Francisco. He recalls her saying, 'There's something a little weird going on here. Would you be interested?' Walker was interested. 'What really got my attention was a man who'd been infected in Boston in 1978, a haemophiliac. He'd gotten really sick, sick enough to be admitted to hospital. Doctors told him he'd have five years to live. But after five years when he was still around, the doctors had to revise their predictions. He said to me, 'I always seem to be at the front end of the curve. Am I still going to die? Do you want to study me?'

At first it was difficult to study these people. All they had in common was that they did not succumb to AIDS and that made them a very mixed bag—some, in fact, did eventually develop AIDS. But in the early 1990s, researchers got a sharper tool with which to probe the non-progressors: a test that measured the amount of virus they were loaded with. The test showed that mixed in with the non-progressors was an extraordinary group of people: the 'controllers'.

Straight after infection with HIV, a person's bloodstream is laden with virus: millions of copies per millilitre of blood. Then as the immune army kicks in, the virus level drops and within a few weeks reaches a steady state of tens to hundreds of thousands of copies of virus per millilitre of blood.[68] This steady state seemed to predict how fast a person would progress to AIDS—the higher the level, the faster their immune system became depleted. People who kept the viral load below 2,000 copies per millilitre of blood were unlikely to progress to AIDS at all. They were dubbed 'controllers'. Some people did even better: their virus level was undetectable. They were dubbed 'elite controllers'. About one in 200 infected people seem to be able to carry out this elite feat.[69]

For Walker the elite controllers were compelling; it was clear they were doing *something* extraordinary to overpower the virus. He studied their blood to get a clue. Some showed what one might expect. When exposed to HIV fragments, their T-cells went ballistic pouring out a chemical alarm called interferon gamma. But in other controllers the T-cells barely raised a whisker. 'What we learned when we studied more elite controllers is that the *strongest and the weakest* T-cells responses are found in this group.'[70]

When I interviewed Walker in mid-2008, he was dismayed by that finding. 'My hunch is that the things we've been looking at as correlates of a successful immune response are not right. We simply don't know what a successful response looks like. We do know that it's *not* neutralising antibodies. We do know that it's *not* T-cells. So the vaccines we have in the pipeline are trying to control things which are probably not that important. It's a part of the problem of science—we've been blinded by a paradigm. We need to step back and start with a clean slate.'

For Walker, stepping back meant taking an unbiased look at the genomes of the controllers. It was not a move without criticism. Like Steve O'Brien before him, Walker's approach was described by some as a 'fishing expedition', an accusation of thoughtless research. (It's a criticism lobbed against much of modern-day genomics.) As Ian Gust put it, 'Just measuring and hoping something will emerge is not likely to be productive. My guess is that these fishing expeditions will generate lots of wonderful papers but no useful product.' Furthermore, there was no certainty that studying the controllers as a group would yield any clear signal. Different controllers might do different things to control the virus. Perhaps these fishermen would return with empty nets.

Walker responded to the criticism this way. 'When I was six my father took my brother and me fishing in a lake in the Rocky Mountains, where we caught nothing. He then loaded us in the car, took us in to town, where there was an overstocked pond teeming with trout, and we both caught fish within seconds. So it really depends on where you fish, and my strong feeling is that the elite controllers, who have been able to do something that makes the virus essentially irrelevant to them, are the right place to fish. To me it is incredible that there is not more focus on these folks. Where else should we be looking?'

Walker also had the advantage of a new generation of tools for genome mining. Twenty years before, Steven O'Brien was limited to prospecting. But since 2005, gene miners have been equipped with machines that can mine the crude ore of thousands of human genomes to extract genes that alter the course of common diseases.

The beauty of this approach is that researchers make no assumptions about the kinds of genes they are looking for.

Completely new understandings were just what the HIV research-ers needed. And it turned out that many of the new generation of gene miners were located a stone's throw from Walker's laboratory in Harvard. At Harvard Medical School, at the Broad Institute in Cambridge and at the Massachusetts Institute of Technology, gene miners had successfully unearthed genes that predispose people to common diseases like diabetes, stroke and rheumatoid arthritis. In 2006, Walker decided it was time for HIV vaccine researchers to take advantage.

***

In the icy spring of April 2008, I immersed myself in the thick of Harvard's genomics cornucopia to attend a conference on HIV and genomics organised by the Harvard University Center for AIDS Research. From the hip affluence of the Broad Institute with its window displays featuring the real-time sequencing of the elephant genome in one window and the isolation of new diabetes genes in another, to the flush of discovery emanating from the young research-ers' faces, there was an exhilarating buzz to the place. Harvard was in the grip of genome fever.

One of those flushed faces was the dynamic Paul de Bakker. With his Eurasian good looks and suave Dutch accent, there's something of the Formula 1 racing car driver about him. The engine he drives can sift through mountains of genome data and pinpoint the genes associated with disease. Technically, it is referred to as a genome-wide association scan. He proved his skill over the last few years at Massachusetts General Hospital, mining genes associated with stroke, diabetes, coeliac disease and rheumatoid arthritis. In 2006, Walker recruited him to drive the HIV controllers study. By early 2008, the study had recruited 1,000 controllers, largely as a result of a huge effort by Florencia Pereyra, another scientist in Walker's lab.[71] When I met de Bakker in April 2008, he was gearing up to compare the genomes of controllers with those of 'progressors', people who progress rapidly to AIDS. What genetic differences would they show? Relishing the task ahead, he told me, 'The beauty of a genome-wide association scan is that there is no bias, no picking of candidates, you let the data lead you.'[72]

***

Like any high-powered engine, the genome mining operations suffered teething problems, mostly related to the statistical analysis. Consider the problem. If you want to test whether a DNA spelling error is associated with a disease, you have to show the association is not just due to chance. Like tossing a coin, the more times you toss, the more likely that something rare will happen by chance like ten heads in a row. So, the statisticians say the more spelling errors you test, the higher you have to raise the hurdle of statistical significance. Once you test a million spelling errors, then the hurdle has to be raised a million times higher![73]

This punishingly high hurdle has led to some major casualties. Few of the candidate genes that Steve O'Brien and others identified in the 1980s and 1990s cleared the hurdle. It is, not surprisingly, a fraught issue. O'Brien has no doubt that statisticians have got things wrong; that they have set the hurdle so high, much of the signal is being left in the slag heap. He told me, 'The statisticians have ruined things. I haven't encountered this much dogma since I was in fourth grade in Catholic School learning the Baltimore catechism.'[74]

Yet in the genome-wide association scans for HIV controllers, a few genes did rise high enough to clear the hurdle.

***

When de Bakker compared the genomes of some one thousand controllers and over two thousand progressors, there were indeed differences. He tested a million spelling errors and 313 cleared the hurdle.
They concentrated in one area of chromosome six. It was the HLA
complex, home of the three genes that make up the uniform of every
cell: HLA A, HLA B and HLA C.

Soaring high above the hurdle in clear first place was a spelling
variation of the B gene, known as HLA B57. In second place, a variant of HLA C also cleared the line. Way, way behind these two in
third place, a tiny signal on chromosome 3 barely cleared the hurdle.
It was in the vicinity of the CCR5 gene.[75]

Although HLA B57 was the clear winner, it was hardly an enthralling finding. The gene's protective role in HIV had already been
fingered by other studies.[76] And, frankly, HLA genes seem to be associated with so many diseases—even schizophrenia.

However, when I spoke by phone with Bruce Walker in March
2011, his tone was victorious. The Walker and de Bakker team had,
for the first time, scanned the genomes of nearly one thousand controllers in a completely unbiased fashion and found that when it
came to overpowering the virus, only the HLA genes were important.
'Out of their three billion letters of DNA, all the hits were in the HLA
genes and nothing anywhere else. It tells us that the heart of control
is the ability of the infected cell to be seen.'

The team had pinpointed exactly which part of the HLA B gene
was important to make bits of the virus visible. The gene forms a
pocket in the HLA uniform for displaying the virus. Just four amino
acids that line the pocket made the difference.[77] The spelling change
near the HLA C gene also seems to help display bits of the virus by
producing more of this HLA C pocket.[78] Having the right HLA B and
HLA C pockets went an impressive 20% of the way to explaining how
much a person could hold back the multiplication of the virus.

There was another reason for Walker's victorious tone. He was
getting very close to a full tactical description of what a successful

battle with HIV looked like. Other Ragon Institute researchers had shown just what these elite HLA B57 controllers did to the virus. When researchers isolated virus from their blood, they were surprised to find it was barely able to multiply. It wasn't that the controllers were lucky enough to be infected with a wimpy virus in the first place. That was clear from their partners who were losing their battle with the virus. No, the controllers had a unique ability to take a robust virus and run it into the ground. And they all ran the virus aground in the same way.

HIV carries only nine genes in its genome. One of these, 'gag', makes a protein that builds a protective inner capsule for the virus's RNA genome.[79] Like the capsule of the moon rocket *Apollo* that protected the astronauts when they catapulted into the Pacific, the virus relies totally on its gag capsule when it is catapulted into the watery interior of the cell. However, when it came to the controllers' viruses, they were getting very little protection from their capsule because it was built from a faulty gag protein.

The best explanation for how this happens is that the controllers' HLA B57 pocket does a stellar job of displaying the gag protein. With gag as conspicuous as a big red handkerchief in the breast pocket of a tuxedo, T-cell soldiers have no difficulty seeing which cells are infected with HIV and eliminating them. To escape detection, the virus is forced to mutate the shape of gag so that it no longer fits the pocket. The odd-shaped gag avoids the pocket but it also makes a defective capsule. The cost of escaping the T-cells is that the once formidable army of virus mutates into a feeble terrorist band holing up in a cave.[80]

After 30 years of searching for an intelligent strategy to fight HIV, it seems Walker has finally hit on one. 'HIV has revealed its vulnerability. Now we know how to make the virus less fit—how to push it into an evolutionary dead end.'[81] But can vaccine developers use this information?

***

Gag may be the Achilles heel of the virus but it's still far from clear how to translate that information into a vaccine. The vaccine used in the failed 2007 trial, for instance, attempted to train T-cells to recognise the gag protein and others. When I asked several HIV researchers how the new information could help to design a better vaccine, they admitted they had little idea.

Training T-cells to recognise bits of the gag protein clearly can't be the whole story either. Only 40% of elite controllers carry the HLA B57 gene, so 60% of them must have different strategies for fighting the virus.[82] There are certainly plenty of hints as to what these other battle strategies might be. Steve O'Brien fingered some of the players years ago.[83] Many immunologists now suspect a different contingent of the immune army, the innate immune system, is also a key player, and its contribution is just starting to be recognised.[84] Another lead comes from an ongoing study of Kenyan prostitutes, some of whom never get infected with HIV. Here, the winning strategy seems to be keeping the immune system very *quiet*. Given that HIV likes to multiply in immune cells, keeping the numbers low might starve the virus.[85]

After 30 years on the trail, the dogged Walker seems quite undaunted by the task that still lies ahead. 'We have a broader agenda. The immune system can do remarkable things; it can carry on surveillance for decades, not just against HIV but also against all the mutations that cause cancers. We want to harness that power. We're just entering a phase where knowledge accumulation is enormous.'

***

PRINCESS MARINA HOSPITAL, GABORONE, BOTSWANA, 29 FEBRUARY 2008.

Eight men to a ward. Most extremely thin—'wasted' is the medical term. They want to live but they are so ill. Daniel Stefanski, the resident specialist, says visiting doctors are taken aback at how sick these patients are. Daniel is an Australian-trained doctor employed by the University of Pennsylvania. Botswana has no medical school so students get medical training overseas, and if they return Daniel joins with local clinicians to teach them the finer points of AIDS medicine.[86]

Daniel is also a friend from Australia[87] and wants to show me the reality of HIV in Botswana. I've donned a white coat for the morning round. His all-female team includes Michelle Morse, an African American medical student from the University of Pennsylvania; Nida Ahmed, a Botswanan doctor of Indian descent, and a nurse whose name I didn't manage to get. I, the new member of the team, am introduced as Dr Finkel, visiting from Melbourne.

Physically speaking, the wards look well-equipped with the disposable sterile plastic and paper accoutrements, mechanised beds, drips and clean blue sheets that you'd see in any Western hospital. There is the same to-ing and fro-ing of nurses and orderlies and stethescoped doctors. What is different are the patients. The ward is a bonanza of medical conditions that few Western doctors would encounter. The poorly defended bodies of AIDS patients are open slather. Some have been invaded by rare things like neurocysticercosis—tapeworm eggs that lodge in the brain. But the spots that show up in brain scans could also be tumours or rampant TB infection. There is also something the doctors are just beginning to recognise: the side effects of HIV drugs. Hollow cheeks and blue fingernails give an outward clue. Internally, the drugs may be causing deadly pancreatitis, kidney disease, heart disease and diabetes.

Finding out just what the patients have is the challenge.

The newly admitted man with sunken, bloodshot eyes is fighting for breath. He sits as upright as he can, straight arms splayed out like struts behind him, totally fixated on the task of drawing air into his lungs. Mostly, it is only the very bottom of the chest that obliges, his diaphragm heaving in and out at an exhausting pace. Is it a 'pulmonary embolism', where a shard from a deep venous blood clot has blocked the lung, or pneumonia, or a mining-related illness, or advanced tuberculosis?

There is the man who presented with nausea and vomiting. He also coughed up blood, right in the face of Dr Ahmed. Besides tuberculosis, they guessed pancreatitis.

There is a white-haired man with lung and brain tumours. He is to be discharged to die at home with steroids to ease his swelling and pain.

There is a severely wasted young man being drip-fed. He has pancreatitis and a bowel obstruction, probably caused by a tumour that they had yet to find.

There is a wasted young man staring vacantly into space. AIDS dementia is the likely diagnosis; a scan showed his brain was shrunken.

Looking very out of place is a nattily dressed, lively young man with a gold earring, sitting atop an unused bed, and looking a little sheepish. Turns out he had multi–drug-resistant tuberculosis, and was supposed to report daily to a clinic for treatment. He had missed two weeks, and was tracked down and escorted to the hospital, though disturbingly not to an isolation ward.

There is a young man in a small room on his own. He has kidney failure, either a result of HIV damage to his kidney cells or a side effect of the drugs. The prognosis for someone with kidney failure is very poor in Botswana, Daniel tells me. They do not have dialysis machines, so patients have to use the do-it-yourself method of peritoneal dialysis—basically washing toxins out of their blood vessels by swishing fluid through the cavity under the ribs. The infection rate is high and patients don't usually last more than six months.

There is a man lying curled up facing the wall. He had come in with cryptococcal meningitis, a treatable infection of the brain's outer membrane usually associated with AIDS. The infection impairs the normal circulation of cerebrospinal fluid, causing very high pressure in the brain, and he needed a lumbar puncture to remove some of that fluid. The hospital had failed to order the correct gauge of needle, so no lumbar puncture could be performed. The ongoing pressure had squeezed the young man's brain and he had gone blind and deaf.

Daniel heaved when he whispered this to me gazing at the wasted young man facing the wall. 'He's only 20. We have to draw letters on his skin to communicate; it's very difficult. He had a great use of language.' There was a blue and red blanket bearing symbols draped over him. Daniel explained it was African witchcraft.

Personally, I had to survive that day in the ward. I was relieved that I felt remote from these poor men, many of whom were in a torpor of weakness, discomfort and pain.

But there was one fellow I did not feel remote from—a child completely out of place in the men's ward. He looked about 11; he was actually 16. His bright intelligence radiated like a beacon. He'd come to Princess Marina a week or so before, stick thin with diarrhoea. But diarrhoea wasn't his problem. Tau[88] was born with HIV; he'd been on drugs, but they were no longer working for him.

Tau's Story

I noticed him on the end bed of one of the men's wards. A tiny figure bent forward, resting his head on a pillow on a table. He looked utterly bored and dismal, turning his head from side to side and staring. Occasionally he picked up a book, but then gave it up. I was just standing around doing nothing trying to keep out of Daniel's hair while he was talking on the phone or attending to something important. I thought I should do something useful for the men on the ward; like help pass the time with a conversation. But I was scared of catching tuberculosis or something, and most of the men did not look like they were up to a chat. Tau might be receptive. I approached him carefully. 'Do you feel like a chat?'

He nodded, his big serious brown eyes in a large lovely child's face, barely looking at me. May I sit here on the bed? 'Yes,' I barely heard again. I picked up the book he had been attempting; it looked like a grade 3 or 4 reader. 'Would you like me to read to you?' 'Yes,' he answered more brightly, but he had a way of not looking at me.

I began reading with all the expression I could muster. And so our relationship began. He was extremely bright, picking up on all the nuances of this really boring book. And he spoke English, beautifully. I visited him several times that day and later we went for a stroll outside the wards together. He had a thirst for knowledge, especially about science. I found myself explaining genetics to him, going back to Mendel the monk who bred peas, and how he realised that there was something that determined whether peas were rough or smooth. By now Tau looked at me a lot.

Daniel found us and said Tau should go back to the ward because Daniel was arranging for him to be transferred to a private ward. Tau seemed very pleased. He gave Daniel a big smile and a hand squeeze— clearly they were mates. I asked Daniel about Tau's prognosis. How could someone so bright and vibrant be so sick? I think I was in denial— this boy had to have a future. Daniel was optimistic, but his expectations were sobering. Tau had lived all his life with AIDS; his mother was dead, and he was cared for by his uncle. He now seemed to be resistant to the drugs, as evidenced by his low CD4 cell count. But Daniel thinks that maybe he has not been given the drugs in an ideal way. Maybe under his care in the private ward, he can garner a better response. While the public ward is free, the private ward costs 400 pula per week (about $66). Daniel is paying this out of his own money. I offered to pay for another month.

About two months after I left, Daniel told me that Tau's diarrhoea returned and he had become depressed. But then he improved. His CD4 count jumped up to 217. Daniel was hopeful this represented a true victory against the virus. He wouldn't know for sure until the level of virus circulating in Tau's blood had been measured.

In August 2009, I caught up with Daniel and he told me Tau had not made it. It was hard for him to talk about it. Checking his notes later he offered: 'Although Tau's CD4 count improved, he did not improve clinically. He continued to be quite sick, didn't gain much weight and had ongoing diarrhoea. A presumptive diagnosis of TB was made and he was started on treatment. I was told that during one admission to hospital, he had felt better and made a decision to stop HAART [the drug treatment] and work closely with a clinical psychologist for emotional support. When he missed an appointment, the psychologist rang the family and was informed that he was very sick at a nearby village. As the family were driving him to the Gaborone hospital, he died en route.'

In March 2011, I asked Daniel what could be drawn from Tau's story. This is what he said.

'I'm not upset about the fact that he died—it was really inevitable and amazing that he survived as long as he did. But the manner was pretty awful. He had such insight, intelligence and dignity about his illness. I remember him saying to me that he just wants to know if the HAART is not going to work, so that he can be prepared for death honestly. He deserved some palliation and home-based care and to die in as much comfort as possible—in his own way.'

For me, Tau's story highlights some key issues about HIV/AIDS in Africa in 2011. First, regarding prevention, there are now some successes. No child should be born with HIV. With three-drug HAART used throughout pregnancy and breastfeeding, Botswana has managed to reduce mother-to-child transmission of HIV to less than 1%. The rollout of HAART in developing countries has been extremely important for all sorts of health, moral, equity, social and economic reasons. There's a long way to go. African patients have proven that they adhere to therapy really well—in most studies, much better than patients in the US and Europe. But to maximise the potential, addressing social stigma, poverty, health system strengthening, nutrition, transport and family support are all important. Tau's story demonstrates this in the starkest way imaginable. And additionally, most counties in sub-Saharan Africa are in much worse shape than Botswana.

***

'Botswana is a small piece of heaven,' said the man slowly in his thick Setswana[89] accent. He had come to fix my door at the Gaborone Sun hotel. Arriving 20 minutes earlier into the reeking hot room, I had slid open the balcony door to find a view onto a deserted car park. Then I could not slide the door back. Even though I was in peaceful Gaborone, capital of Botswana, spending the previous two weeks in crime-ridden South Africa had unnerved me. A determined intruder could make it in, even if I was on the first floor. Best to get it fixed. The hotel repairman had come promptly and taken about 20 minutes to fix it. I thanked him, saying I would be able to sleep peacefully as a result. He looked at me for a moment with an injured expression and then retorted, 'You would have slept peacefully even with a broken door.' Then he made his proclamation about the heaven that was Botswana.

'A small piece of heaven' certainly fits the image of Botswana I had gleaned from reading Alexander McCall Smith's *The No. 1 Ladies' Detective Agency*, part of my background research. Botswana also holds the dubious distinction of being the best place in the world to study HIV. In 1990, HIV was virtually unknown there.[90] The country was rapidly modernising thanks to the dual blessings of newly discovered diamond mines and a stable, democratic government. Its residents enjoyed the highest life expectancy of anywhere in Africa, 65 years.[91] In a continent ravaged by war, corrupt governments and starvation, Botswana truly was a 'small piece of heaven'. But through the 1990s, HIV spread like wildfire, especially ravaging women. The average life expectancy crashed to 40 years. Festus Moggae, the Prime Minister, warned, 'Our population is threatened with extinction. People are dying in chilling numbers'.[92] In stark contrast to then South African president Thabo Mbeki's denial of HIV as the cause of AIDS, in 2002 the Botswana government, with support from the Gates Foundation and the pharmaceutical company Merck, rolled out free antiretroviral drugs to the needy population. It was the first country in the world to do so.

For Max Essex, Botswana was the obvious place to look for an answer to the question that has compelled him for decades. While the virus has spread throughout the world, there is nowhere outside of Africa where adult infection rates exceed 4%. In Europe, the US, Australia, China and India, less than 1% of the population carry the virus. And within Africa, southern Africa is by far the worst afflicted. While northern Africans have a prevalence of 0.4%—the same or lower than their European neighbours—and middle African countries average 4.5%; in southern Africa the figure averages 23% for the countries of Botswana, Lesotho, South Africa and Swaziland.[93] In the antenatal clinics of Botswana, a third of women on average are infected but in Francistown, the second largest town, the figure is an apocalyptic 40%.[94]

Some experts have put it down to an unfortunate clustering of high-risk factors. There is, for instance, a culture of having multiple concurrent partners, which spreads the virus like a chain reaction. In

Botswana men often have a large house for the main wife and smaller ones for their mistresses.[95] Others point to high-risk practices like anal sex and dry sex, or the prevalence of other sexually transmitted diseases that pave the way for HIV. By contrast, northern Africans may be protected by Islamic sexual mores or ritual male circumcision, which can reduce the infection rate by 60%.[96]

Essex has never been convinced by these explanations. 'It sort of slaps you in the face that rates are a lot higher in Africa and much higher again in southern Africa. You can't get very far arguing cultural differences or poverty. Papua New Guinea and Cambodia [with a prevalence of 0.9% and 0.5%] are every bit as poor as southern Africa. Thailand has a huge sex industry [and a prevalence of 1.3%]. I think there are only two reasonable explanations. One is the genetics of the virus; the other is the genetics of the people. In southern Africa both are different—I suspect both are important.'[97]

In the icy spring of 2008, I visited Essex in his office at the Harvard School of Public Health, a grey stone building that stretches for an entire block. Essex is a courteous, serious man in his late sixties. After 25 years on the campaign against HIV, clearly his pace or sense of urgency have not slackened. I managed to squeeze in an interview with him before his early morning conference call to Africa.

Essex trained as a vet and is one of the pioneers of HIV research. Together with Robert Gallo and Luc Montagnier he received a Lasker award in 1986, second only to the Nobel Prize in scientific prestige, for his contribution to identifying HIV as the cause of AIDS.[98] Essex has continued to sleuth HIV over two and a half decades and much of the sleuthing has been directed at the mystery of African AIDS. Why did it spread so much more rampantly into the southernmost countries? And why did it infect more women than men here, while doing the reverse in America and Europe? To answer those questions and others, in 1996 Essex set up Harvard Botswana, an HIV research institute on the grounds of Princess Marina Hospital.

Perhaps the virus itself was different? In other words, some countries were just unlucky enough to be infected with particularly virulent strains of the virus or strains that preferred women to men. Strains of HIV do indeed differ. This was dramatically illustrated in 1986, when researchers discovered that the populations of West Africa were infected with a much more benign type of virus. Only 1–2% of the population were infected and of those only 1–2% went on to develop AIDS. The virus was so genetically different that it was labelled HIV type 2, or HIV-2.[99] The discovery of this virus made Essex wonder if more subtle genetic differences could explain the different characters of HIV. To a tantalising extent they did. Within the main virus type, HIV 1, there are different 'subtypes'.[100] Subtype B is found in those countries where the epidemic predominates in the gay and drug-using population, and also in the white population of South Africa. Subtype C is found where the epidemic is largely heterosexual, like southern Africa and India. Subtype E is closely related to C and has fuelled the heterosexual epidemic in Thailand.

Laboratory tests by Essex and colleagues shed some light on why the different strains behave differently.[101] Those associated with heterosexual spread, C and E, are good at entering and growing in the cells that line the vagina and foreskin—so-called Langerhans cells (this explains why male circumcision protects against HIV infection). By contrast, subtype B doesn't do that well in Langerhans cells, which probably explains its very low rate of heterosexual transmission. Studies suggest it's as low as one in 1,000 exposures of heterosexual intercourse.[102] To enter the body, subtype B probably relies on ruptures to the tissue lining, more common with anal intercourse.

Notwithstanding his own compelling discoveries, Essex doesn't believe viral subtypes are the whole answer. For one thing, it is hard to understand why subtypes of HIV seem to stay confined in one population when the virus has clearly demonstrated an impressive ability to traipse around the world. HIV-1 seems to have originated in equatorial Africa in the late 19th century when it jumped from chimp to human populations. From there it spread to Haiti, then to the US and back to Africa and the rest of the world.[103]

***

When it comes to human genetic variation, nowhere is it greater than across Africa.[104]

This is where the family of man began some 200,000 years ago. The diverse races of Africa represent thick gnarled branches growing out in different directions from the trunk of the family tree. They have had hundreds of thousands of years to diversify. By contrast, the people who populate the rest of the world are all recent offshoots of one branch. They show far less genetic diversity.[105]

So it is not hard to imagine that the genes of Africans might be very different from those of Europeans or Asians when it comes to susceptibility to HIV. Small-scale studies from Harvard Botswana and others have already shown that to be true. For instance, the CCR5 Δ32 gene variant that Steve O'Brien found to reside in about 10% of Europeans is entirely absent from the African population. Africans also miss out on the HLA C protective variant and another one called ZNRD1. And it turns out that a gene variation that protects them from malaria—DARC—may predispose them to HIV infection.[106]

Given these inklings of difference, Essex told me, 'I think it would be derelict not to examine the genetics of the people.' His notion that genes hold part of the answer to AIDS in Africa has already started to provide some dividends for people in Botswana, as I discovered on my visit to Harvard Botswana in February of 2008.

***

Dr Bill Wester met me in the bustling front grounds of Princess Marina Hospital and we made the short stroll around the back to

Harvard Botswana. Seeing its granite façade rising above the low cream brick buildings of the hospital, the metaphor of an ivory tower was inescapable.

Wester is a genial, softly spoken man. He and his wife came to Gaborone in 2001 to do AIDS research and helped set up *Masa* or 'new dawn'—the program to deliver HAART. There was no guarantee that the AIDS drugs would produce exactly the same effects in southern Africans, nor the same side effects. Wester found a case in point, a type of drug known as a nucleoside reverse transcriptase inhibitor (NRTI). The drug puts a spanner in the works of a crucial bit of the virus machinery, the reverse transcriptase that rewrites the virus's RNA genome as DNA. If the virus can't be rewritten as DNA, it can't replicate. But in some people the spanner also creates havoc in their mitochondria. Mitochondria are power plants that live inside our cells and generate energy from sugars. Curiously they carry their own DNA, probably because they evolved from bacteria. The enzyme that copies their DNA is also susceptible to the NRTI drug.[107] The drug can damage the mitochondria's DNA and when it does, power generation fails and nasty by-products like lactic acid build up—the same stuff that makes your muscles ache after a workout because your mitochondria can't keep up. Ultimately it leads to lactic acidosis, a condition that acidifies the blood and causes death within days.

In studies on Western populations, this side effect is seen with a particular class of NRTIs, known as D drugs, in 0.1–0.4% of people. But in the population of 3,000 Botswana, people that Wester has been studying, 1–1.2% are showing the side effects, up to 12 times the rate. And whereas in the West the symptoms appear in men and women equally, in Botswana women are overwhelmingly the victims.[108] The first symptom is a deceptive one: women start putting on weight, usually a sign that people are responding well to drug therapy. But these weight-gainers are the ones who suddenly develop lactic acidosis. If not taken off their medication promptly they proceed irreversibly to death. Thanks to Wester's findings, cheaper D drugs are being phased out in Botswana. 'The good news is our research paper changed public policy.'

The different responses to drugs are just one more compelling reason to peer into the genomes of Africans. While many studies have looked at how the genes of Europeans predispose them to HIV, the question hasn't been addressed in the population most afflicted by HIV.[109] At Harvard Botswana, Essex is collaborating with Steve O'Brien to do the first systematic study of HIV modifying genes on a southern African population. O'Brien, compelled as ever by the secrets lurking in genomes, told me, 'Not everyone agrees with Max's hunch. But I'd love to find the genes that predispose to heterosexual transmission.' The pair will try to identify genes that predispose people, particularly women, to infection; genes that determine how fast people progress towards AIDS; and genes that determine how fast people recover after drug therapy.

***

Despite endless prevention campaigns like Botswana's ABC: 'Abstain, Be faithful, Condomize', HIV continues to spread across sub-Saharan Africa, most aggressively in young women. The majority of those now infected are likely to die terrible deaths because, unlike Botswana, they don't have access to treatment;[110] while those fortunate enough to be treated may end up trading one disease for another. There is no vaccine and no hope of one for at least a decade.[111] And researchers cannot even say why it has spread so rampantly in some places and not others.

In places like sub-Saharan Africa it seems we are losing the battle against HIV. But we may yet win the war. HIV has taught us to look for answers within. Instructed by genes from rare individuals, the immunologists are learning new lessons that will teach them how to take command of the immune army. The rewards of that will be

profound, not only for tackling HIV but also malaria and tuberculosis and the plethora of diseases from cancer to arthritis, that at some
level result from an immune system gone awry. The next plague virus
that appears out of nowhere will find the human race considerably
better prepared. For now, HIV needs to keep blipping loudly on the
global radar screen. By the time this book is published some 33 million people will have died from AIDS.[112]

6

Feeding Nine Billion

On 17 March 2009, I met a legend: Norman Borlaug, the father of the Green Revolution. This revolutionary never wore a black beret or fatigues, nor did he ever throw a Molotov cocktail. Borlaug was a wheat breeder. His tool of revolution was phenomenally productive wheat that doubled yields. Rice breeders copied his methods to achieve similarly fabulous gains. And from the mid-1960s, the bounteous breeds of wheat and rice spread across Asia and then the world in what became known as the Green Revolution.[1] This revolution overturned a chilling prophecy first espoused by Thomas Malthus in 1798 and again by Paul Ehrlich in 1968: that the human population would outgrow its ability to feed itself.[2]

Borlaug's Green Revolution earned him the Nobel Peace Prize for saving hundreds of millions of people from death by starvation, and the epithet, 'The man who fed the world.' Indeed, Borlaug fed the world for decades. The enormous yield jumps meant that by the mid-1980s, the world supply of wheat and rice leapfrogged ahead of population growth. Europe acquired food mountains that stored months' worth of grain.[3] Globally, food became cheap.

In 2008, everything changed. The mountains shrank to a 30-day supply. Food prices soared: rice went from US$400 per ton in January 2008 to more than $1,000 per ton in May.[4] A billion people who lived on a dollar a day could now afford to eat only once daily.[5] Dozens of countries saw food riots. What happened? Global grain production had been hammered by a slew of factors: droughts in Australia, floods in Asia, biofuels displacing food crops, and rising prices for fuel and fertiliser. Then, topping off the sense of apocalypse, a vanquished foe of wheat had risen from the dead—stem rust, the most feared of all wheat diseases.[6] It was a perfect storm and the sailors were caught unprepared. Because underpinning it all, Malthus's prophecy had indeed come to pass: agriculture had lost the race with population growth. The world population was growing at 1.4% while the yield increases for the world's two staple grains, wheat and rice,[7] had flattened out at around 0.5% for wheat and 0.9% for rice.[8] Why did it happen? At least partly, because the decades of plenty had left global agriculture complacent.[9] The international network that had fomented the Green Revolution had grown threadbare and toothless.

That was why I got to meet Norman Borlaug. Aged 95, he was the keynote speaker at a wheat conference held in Obregon, Mexico, the birthplace of the Green Revolution. Here in 1944, the Iowa-born Borlaug had been hired by the Rockefeller Foundation to help Mexican farmers fight stem rust. The wheat he bred not only resisted rust and other diseases, it was super wheat. It was short and stocky, so it concentrated its resources into the grain rather than the stalk—so doubling yields! And it could easily adapt to growing in different parts of the world. Knee-high Green Revolution wheat now stretches in a golden sea from Azerbaijan to Australia.

Borlaug won these victories in his youth. But now his nemeses were back: wheat rust and flat yields. Like Beowulf, the aged heroic knight of the old English poem, Borlaug heard the call to arms and donned his creaky armour to lead the campaign for wheat once more.

My first glimpse of the legend was at the opening session. Seated in the centre of the front podium behind a white tablecloth and surrounded by the other keynote speakers, Borlaug looked old and frail as the crowd of photographers snapped and flashed incessantly at him. But when he delivered the opening speech willed through bursts of his creaky Midwest-accented voice, there was no doubt. This was the legendary Borlaug, great not just because of his skill as a breeder but great because of his humanism and his dogged determination that got past the political and bureaucratic roadblocks that stood in the way of feeding the world.[10]

His speech was ultimately a plea to revive the internationalism of agriculture. He recounted the beginnings of the Green Revolution in Obregon more than 60 years ago when his organisation, supported by the Rockefeller Foundation and the Mexican government, had created a 'United Nations' of agriculture. He recalled how the farmers and scientists of the Yaqui Valley had welcomed people of all languages, races and colours as they came there to learn the ways of wheat. Faltering with emotion, he intoned, 'That was the lesson they learned here.'

After Borlaug's address, the snapping photographers disappeared. Borlaug, in his wheelchair, relocated to the front row of the audience to hear the scientific presentations. The impenetrable cordon around him had vanished and when the session ended he sat there alone and available. I didn't dare to approach him cold. I saw his daughter Jeanie speaking to another person and enquired whether I might ask Borlaug a question. Enthusiastically, she obliged. 'Daddy, this lady journalist from Australia would like to ask you a question.'

Borlaug fixed a pair of keen, blue eyes upon me. I had a respectable question for him, I thought. Borlaug's greatness as a breeder lay in the fact that long before DNA was even part of the vocabulary, he had managed to generate extremely high-yielding wheat that was effectively universal because it could grow in many different environments. It also carried a winning set of disease resistance genes that protected wheat from rust and other pathogens. The royal flush had been so effective it had protected wheat from stem rust for 50 years until the recent outbreak. 'How did you do it?', I asked him. For a split second he processed my question, then his answer shot out carried more by breath than vocal cords, 'Feel; I did it by feel.' It was accompanied by a raised, determined right hand with fingers splayed to show me how.

Next day, during a tour of the fields in the warm Mexican sunshine, I revelled in seeing the wheats I had read so much about. I never tire of looking at wheat. Erect and golden, the grain in the thick ears is arranged in a pretty herringbone pattern and its grassy leaves dip and curl so gracefully.

Each variety is different—height, shade of green, sheen, curl of the leaves, length of the awns (the long hairs extending from the top and sides of the wheat ear), thickness of the ears, sooty markings on the stalk and leaves … Whatever I can rattle off is just the beginning;

Wheat, by Christine Zavod

those who have 'feel' see much more. There, amid the plots, Borlaug's heirs from CIMMYT, the international wheat- and maize-breeding centre that grew out of the Rockefeller-funded institute, gave tutorials about the varieties growing in front of us.

Borlaug in his wheelchair, wearing a tan hat against the hot Mexican sun inscribed with the words Borlaug Global Rust Initiative, and flanked by a cordon of devotees, peered occasionally from under his hat at the wheat and listened attentively. It was a picture worth a thousand words. The faces—brown, black, white and every shade in between—of those who stood protectively around him were the faces of people he had trained decades ago in this United Nations of Obregon. They had gone back to Pakistan, India, South America and Africa to carry on the Borlaug credo of wheat-breeding.

At one station, the person giving the tutorial was Dr Julio Huerta, a stocky middle-aged Mexican plant pathologist, who decades before had been Borlaug's technician. After the rest of the group moved on, I asked Huerta my question again. 'How did you do it?' He took me into

Borlaug surrounded by wheat breeders. (Photo: E. Finkel)

the field to give me a first-hand idea of 'feel'. He grasped one of the plants near the bottom and showed me how to feel for the number of tillers (stalks) it has and how sturdy it is; and how to look at the ears and leaves. You scan the rusty pustules and flecks on the leaf reading them like Braille; they tell you how much resistance the plant carries to rust and other diseases. You look for a bit of blackening on the seed coats (glumes) and stems. That's good; it's a hallmark of a rust-resistant plant. You look for white tips on the leaves, again a hallmark of rust-resistance. It's sun-up to sun-down work, done as fast as possible to compare the plants accurately. Tens of thousands of plants had to be screened this way in a season, he told me pointing out over the vast field, to find the few that had the royal flush of genes for both resistance and yield.

'Borlaug was a taskmaster,' Huerta said, punching his fist to illustrate the fact. 'He would say to me, "I want more of that black stuff." He was tough but he had to be. Now I'm the same to get through all this wheat.'

All that wheat to screen. (Photo: E. Finkel)

Feel and brute force was how Borlaug and Huerta did it. Today, plant-breeders need to repeat the feat. The United Nations Food and Agriculture Organization calculates that to feed a population of nine billion by 2050, breeders need to raise wheat and rice yields by 70%.[11] They also need to give wheat robust protection from stem rust while maintaining its ability to thrive in diverse environments. But the methods used by Borlaug are no longer delivering. Since 1995 wheat and rice yields have been almost flat.

Borlaug achieved his monumental gains by changing the architecture of the wheat plant. The traditional Mexican wheat he encountered in the fields was the height of a man's shoulder and spindly. Once fertiliser and irrigation fattened the grain, the plants became top-heavy and keeled over. To fix the problem, Borlaug crossed them to a dwarf Japanese wheat called Norin 10. The hybrid 'semi-dwarf' plants weren't just more sturdy. Marvellously, all the resources previously bound for the stem and leaf now concentrated in the grain. That was how yields doubled and sparked the Green Revolution. Afterwards, through the 1980s and 1990s, breeders kept tinkering with plant architecture, shortening the stalks or finding ways to increase the number of ears borne by each plant. They managed to eke out 2–3% yield increases each year till the 1990s, when they hit a plateau. Wheat plants seem to have gone as far as they can shrinking their stalks to divert more than half their resources to grain;[12] rice plants seem to have produced more ears than they can fill.[13]

Tinkering with the plant's chassis won't do the trick any more. What's needed is to go inside and tinker with its photosynthetic engine. This is the engine on which all life depends; every atom of oxygen we breathe, every molecule of carbon in our bodies, was produced by photosynthesis. Fuelled by the sun's light it knits carbon dioxide and water into sugars and releases oxygen as a by-product.

A tooled-up new breed of international researchers has already lifted the plant engines onto the hoist and are hammering and soldering away; their audacious goal is a complete overhaul. The International C_4

Rice Consortium is doing it for rice and the Wheat Yield Consortium is doing it for wheat.

These heirs of Borlaug are thinking big. Imagine a rice plant producing ears closer to the size of corn without increasing its use of water or nitrogen. It's not so far fetched. Corn is powered by a turbocharged engine that evolved 30 million years ago during a phase when carbon dioxide concentrations suddenly plummeted. Sixty-two different plant species did it through chance.[14] Surely the plant engineers should be able to do it—retrofit a corn-like engine into rice—much faster.

* * *

When our planet was still a baby, barely a billion and a half years old, purple-green bacteria built themselves a remarkable machine to harvest the radiant energy of the sun and store it chemically as a sugar molecule.[15] The core of this machine is the enzyme Rubisco. Rubisco's job is to snatch carbon dioxide from the air and knit it into a stable energy-rich chain of molecules that we know as sugar. The engine that drives this knitting machine is solar-powered. Photons of sunlight split water into protons and electrons that in turn drive the synthesis of the high-energy molecules ATP and NADPH, which act like batteries to power Rubisco.

The carbon dioxide concentration of the atmosphere a billion and a half years ago was about a hundred times higher than today,[16] not great for oxygen breathers but heavenly for the purple-green bacteria. Their Rubisco machine guzzled the carbon dioxide and churned out sugars. Photosynthetic life went from strength to strength: from purple-green bacteria to algae, to mosses and ferns, cone-bearing plants, and ultimately the flowering species. They and their Rubisco engines did so well at sucking carbon dioxide out of the atmosphere and knitting it into sugars (that in turn were knitted into wood, coal, animal skeletons and coral reefs) that carbon dioxide levels gradually fell.

However, 30 to 40 million years ago, carbon dioxide levels fell rather suddenly—from about 1,000 parts per million to 300 parts per million (lower than the level we have now of 390 parts per million).[17] Plant engines that had been used to guzzling carbon dioxide suddenly found themselves spluttering and at risk of a stall. It was a huge stress. Under that selective pressure, some 62 species of plants evolved a new way to thrive.[18] Their adaptation was to pump up the carbon dioxide inside the cell, raising the internal pressure tenfold, rather like the way a turbo-charger increases engine performance by delivering a higher concentration of volatile fuels. Different plants

A turbocharged engine – the C_4 plant

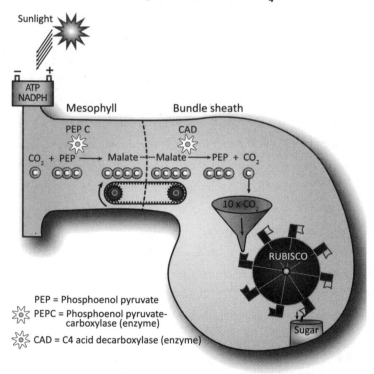

Powered by solar energy, mesophyll cells capture CO_2 and store it in the four-carbon chain, malate. Malate is ferried into bundle sheath cells where the CO_2 is unpicked from the chain, raising the concentration ten-fold and turbo-charging the sugar-making Rubisco engine.

managed it in different ways, but they converged on one step of the strategy. They all knit carbon dioxide molecules into a four-carbon storage chain[19]—the reason they are called C_4 plants—and ferried it into a gas-tight chamber where the carbon dioxide was released.[20]

The successful C_4 plants with their turbo-charged Rubisco engine are quite obvious today. Think: corn as high as an elephant's eye, sugar cane or bulrush millet—12 weeks and you need a ladder to see over the top of it. King of the C_4 plants is elephant grass, delivering 88 tons of dry biomass per hectare—three times the performance of wheat at its very best![21] But C_4 plants can produce their largesse on relatively poor soils and use much less nitrogen and water than their C_3 cousins.[22] It's no wonder C_4 plants are commanding attention as a source of biofuels. But since 2008, another group has been eyeing C_4 plants with a different goal in mind. Plant engineers funded with US$12 million from the Gates Foundation have set out to try and retrofit the C_4 engine into rice. It was John Sheehy's idea. The head of rice physiology at the International Rice Research Institute (IRRI) in The Philippines saw yields bottoming out and new breeds of rice producing more ears than they could fill.[23] It was time to try and boost the engine. To begin, Sheehy assembled a team of engineers. It wasn't hard; they were a small, unfashionable group of academics from the UK, Germany, Canada, America and Australia, who all knew each other on a first name basis; all diehard aficionados of C_4 photosynthesis.[24] Robert Furbank was one of them. Besides being an incurable tinkerer, he is the director of a state-of-the-art facility that can measure the workings of the plant engine: the High Resolution Plant Phenomics Centre at CSIRO, Canberra, Australia. It is a facility that has taken 'feel' from being an art to a science.

**

The plant engineers have the hubris to try their wild idea because of two new pieces of 'omics' technology. 'Omics' is a distinctly 21st century suffix. It means doing things en masse. Genomics enables researchers to quickly scan through lots of genomes. Phenomics allows them to scan the phenotype (the characteristics) of lots of individuals. Borlaug and Huerto, who felt-up hundreds of plants a day, almost qualified. In Furbank's centre they scan thousands a day. But it's not just about numbers. Phenomics also looks far more deeply into the inner workings of the plant. 'Feel' was about guessing what was going on in the plant from outward hints rather like a doctor guessing at what's going on with the heart using a stethoscope. The modern day phenomics people have the equivalent of MRI and CAT scans.

Genomics combined with phenomics has empowered plant engineers to try and rebuild the plant photosynthesis engine. A genome provides the parts list of a plant. But on its own, a genome is not much good for redesigning the plant engine. Imagine receiving the parts list of a VW beetle. You see tens of thousands of listed components, only some of which you recognise—perhaps the words 'cylinders' and 'fuel tank'. Now you are asked to redesign that beetle to perform like a Ferrari. Fine: you start tinkering. Let's try and add a few more cylinders and increase the size of the fuel tank and stretch out the shape of the chassis while we're at it. Of course, what works on paper may not work at all in a real car. Extra cylinders might just blow up the car, or they might not connect with the rest of the engine. As any engineer will tell you, you can't redesign a car or make any advances in engineering without a means of measuring each individual change. To re-engineer a car, you need a dynamometer, a diagnostic machine upon which you hoist the car to measure the performance of the engine components. To re-engineer plants, you need phenomics.

***

You wouldn't pick Bob Furbank as being a plant scientist of international renown—maybe a high school football coach. He's a dark-haired, pleasant-looking man in his early fifties, with a sort of hungry look about him. The football coach comes from the fact that he's blokey and approachable and has an easy way with words. In the same jaunty manner a coach might use to tell you how to place the ball, he is describing, you realise with a surprise, the intricacies of how to re-engineer a rice plant. The blokey vernacular comes from his origins in the steelworks town of Wollongong. That's also where he gets his mechanical flair: he grew up pulling engines apart. The son of a steelworker, Furbank was the first of his family to go to university. He might have done law at the University of Sydney except that he'd been totally captivated by photosynthesis. 'I remember a physics high school teacher describing it and thinking, "This underpins all life on earth."' Maybe it was also something to do with the fact that it was another engine to pull apart. So he turned his back on law for science at Wollongong University. A lecturer saw his potential and steered him towards a PhD at the Australian National University, where he began his dream in earnest—taking apart the engine of a turbo-charged C_4 plant. One of his three supervisors was Hal Hatch,[25] probably Australia's most famous plant scientist, who won the 1991 International Prize for Biology for discovering the enzymes that are crucial to the C_4 turbo-charger.

On his first day, Furbank recalls that Murray Badger, one of his PhD supervisors, presented him with a broken mass spectrometer (a machine for measuring minuscule quantities of different molecules based on their mass) and a manual and said, 'Before you can start you'll have to fix this.' Furbank delicately unscrewed four pins and triggered an avalanche of components onto his desk ... It took him about six weeks to put the machine back together again. Three years on and he made some impressive discoveries about the workings of the C_4 machine. One of the reasons it was so efficient was that it prevented Rubisco from engaging in some wasteful chemistry. Rubisco is supposed to grab CO_2 molecules and knit them into a sugar chain. But occasionally it grabs an oxygen molecule and knits that into the chain instead. The end result is to break rather than make the sugar chain, and to squander a molecule of CO_2. (It's called photorespiration.) C_4 plants douse Rubisco with such high levels of CO_2 that a stray oxygen molecule is unlikely to be knitted into the chain.

In time-honoured tradition, Furbank went overseas for post-doctoral study, coincidentally to another town famous for its steelworks, Sheffield in the UK. Here he was meant to continue his studies of C_4 plants but as he found, 'the dim dark reaches of Yorkshire were somewhat of a challenge'. C_4 plants are not designed to grow in dim, dark places. So he trained his attention on the standard C_3 photosynthetic engine. Researchers traditionally divided its operations in two. The front end harvested light to charge up the molecular battery. The back end, Rubisco, used the battery power to knit carbon dioxide into sugars. Like two gears of an engine, these reactions had to be coupled in some way, but few researchers had probed at how these so-called *light* and *dark* reactions meshed. Furbank found an answer. They did indeed mesh, but not tightly as in gears connected by a fanbelt. Rather, Furbank describes it as 'viscous coupling', the way two gears are sometimes loosely coupled by a layer of thick oil. If the Rubisco gear could not keep up with the energy spinning out of the light-harvesting gear, it needed a way to turn down the intensity of light being harvested. Plants can't get out of the sun, so Furbank found that they match their light harvest to their energy needs by shading their solar panels,[26] smearing them with xanthophylls, the plant's equivalent of a sunscreen.

Furbank's time in Sheffield was extremely productive. In 1987, he returned to CSIRO Plant Industry in Canberra with laurels, having won a prestigious QE2 fellowship that supported his research for three years. He had the feeling 'that he was being groomed by Hatch as a successor'. He also had the freedom to work on anything he wanted. He chose to systematically take apart the C_4 engine. His crucial partner in the work was a lady with golden hands, Julie Chitty, also his wife-to-be.

Chitty developed a model system using the obliging C_4 plant *Flaveria*. The idea was to throw spanners in the works of the different parts of the C_4 engine by disabling genes for enzymes. While Chitty made plants with broken enzymes, Furbank measured the effects on the plant engine using techniques perfected during his PhD and at Sheffield. By the late 1990s, their elegant efforts bore fruit and they were churning out papers on how the C_4 engine was regulated, papers Furbank admits to 'being very proud of'. It should have positioned Furbank as a star in the firmament of global plant research.

It did not.

Apart from fellow aficionados of photosynthesis, the world was not really interested. In particular, CSIRO was not interested. They were the national research institute with a remit to pursue problems of practical significance to Australian agriculture, especially as more and more of their funding had to be sourced from industry. In the case of CSIRO Plant Industry that meant focusing on research of importance to agriculture, and in the late 1990s with food mountains towering in Europe, no-one could see the need to go pulling apart plant photosynthetic engines. Particularly those of the C_4 variety. Australia's main cash crop was wheat, a C_3 plant.

Overnight, Furbank had to abandon his research into photosynthesis. 'I was told, "Photosynthesis is not a viable area to work in. It's your job to work out how to refocus your activity."' That probably explains the hungry, even slightly aggrieved look that still flickers below the surface of Furbank's jaunty demeanour. He was not aggrieved enough to leave CSIRO Plant Industry, even though he'd had a job offer from Cambridge. 'I always wanted my research to have an impact on crops and CSIRO Plant Industry is one of the few places in the world where you can do world class fundamental research but rub shoulders with scientists testing crops in the field.'

***

It wasn't just photosynthesis research that fell on hard times. Globally, institutions were divesting themselves of agricultural research.[27] As far as yield was concerned, there was no need for more research. The Green Revolution had solved the problem.

At CSIRO Plant Industry the focus moved to helping plants cope with stresses like drought and salty soil. And to producing nutritious grains with health benefits such as high fibre barley to protect against colon cancer or canola enriched in omega-3 fatty acids to protect against heart disease.[28] As Furbank put it, 'The question became, "What else can we do with food?"'

So Furbank refocused his thoughts from the intricacies of plant engines to the needs of crops growing in the field. It was the research of his CSIRO colleague Richard Richards[29]—a virtuoso wheat breeder who was making long sought-after gains in drought tolerance by thinking laterally about the character of wheat—who offered him a bridge.

Australian wheat is typically planted in the beginning of winter to tee up with the rainy season. Problem is, in the spring the soil starts drying out, yet the grain developing inside the stem still needs three months to fill with carbohydrates—a difficult task with limited water. Richards found that some varieties of wheat did a better job of fattening their grain because they had stored sugar in their stems during the winter. Sugar storage was close enough to photosynthesis to get Furbank interested. He put together a team of scientists to develop wheat varieties that were adept at storing sugar in their stems over the winter. The first step was to hunt for the genes involved in storing sugar.

Sometimes genes are easy to hunt down. For instance, a hunt for the 'dwarfing' genes that led to the dwarf wheat of the Green Revolution unearthed two genes.[30] And the hunt for the gene that stimulated rice to produce more ears nailed just one gene.[31] These genes were easy to track down because just one or two powerful genes had caused the trait. What's difficult is when a trait is caused by lots of genes working together, and each individually has a small effect. Like a royal flush in a poker game, the individual cards don't have much power; it's the set that counts.

The plant scientists suspected that the stem sugar trait would be of the latter variety—lots of genes and very hard to find. But they were emboldened by the new technique of 'genotyping' using gene markers that promised to make their job easier.[32]

***

Finding the genes required crossing a parent plant that had lots of stem sugar with a parent that had little stem sugar. The offspring of the cross inherited different combinations of their parents' genes. It was just a matter of checking each seedling to see which had the best stem sugar stores and cross-checking to see which gene markers they inherited. In other words, matching their 'phenotypes' with their 'genotypes'. However, while genotyping was a 21st century technology, phenotyping was stuck in the Stone Age.

Furbank vividly recalls his first few days in Ginninderra field station 20 minutes outside Canberra in 2002, grubbing around in the dirt with a steak knife bent at 90 degrees like a sickle, cutting stems and leaves off thousands of wheat plants. Then back in the lab all those samples had to be ground up and their sugar content measured in absorption spectrometers that read the wavelength signature of sugars, rather like the way astronomers detect the signatures of different molecules in stars.

Three years later the genotypes and phenotypes were cross-checked. And the researchers found *nothing*. There was no clear set of genes that corresponded to wheat's ability to store sugar in the stem. Clearly, it was a very complex trait—perhaps hundreds of genes were involved. Tracking such plants by their genes might not even be possible. Grinding up their stems was not the answer; not only was it ridiculously time-consuming, it destroyed the plant you might want to breed from. Breeders needed another way of tracking the winners.

In his mind, Furbank started fantasising something like the 'tricorder' from *Star Trek*, a little scanner you could wave over a plant and read the wavelength signature of the sugars in different parts of it without harming it. If breeders had access to that, they could use it to quickly and cheaply pick the best performing plants to breed from.

***

CSIRO Plant Industry nestles on the side of Black Mountain about ten minutes' drive from the centre of Canberra. The grounds are pleasantly treed and shrubby. Low-slung rectangular buildings like army barracks are the labs, and just behind them are rows of quaint glasshouses growing different varieties of wheat, barley, canola, cowpeas, cotton and other things. Strolling around in the brilliant dappled Canberra sunshine, the eucalyptus-scented air is invigorating and parrots screech and currawongs chortle from Black Mountain's forested flank.

It's a mellow, utilitarian 1960s sort of place, except for two buildings. One is the glass and steel 'Discovery Centre' which is part museum, part zoo. A visitor entering through the grand walkway and glass doors is drawn to cabinets and displays exhibiting the latest in plant research but if they glance to the left, they find themselves peering through glass panels at enclosures with living specimens: white-coated researchers at work in their laboratories.

The very newest building in the compound was opened in August 2009. It is the High Resolution Plant Phenomics Centre. As you enter, you are met by bare grey letters on a white wall that read, CHERISH THE EARTH FOR MAN WILL LIVE BY IT FOREVER.

The ante-room is all soft pastels and rounded surfaces, and you're welcomed by an overhead flat-screen display. It's futuristic in a retro sort of way. Furbank describes it as 'Thunderbirds style' and explains it as the architect's homage to the original 1960s building. That was the 'Phytotron', which housed the start-of-the art technology of its time— air conditioning—allowing researchers to grow plants at controlled temperatures. The entire ground floor had housed a pool for that purpose. The High Resolution Plant Phenomics Centre has been built to capture the state-of-the-art agricultural technology of the 21st century. Furbank is not just calling it phenomics anymore; it's got a sexier name: digital agriculture.

As a visitor to the centre soon discovers, it's an apt description. All the fuzzy stuff that was once embodied by 'feel' or 'phenotyping', is being measured here with digital precision, and fed straight into computers generating terabytes of data per day! As I tour around I get the feeling that the place is filled with boys' toys. One of the 'toys' graphically demonstrates what 'digital agriculture' is all about. Furbank and his right-hand man Xavier Sirault, whose official title is Engineer Scientist in Plant Phenomics, developed this beauty after a trip to Fox studios in Sydney in 2007. The LIDAR[33] machine they saw there digitised people, transforming ordinary (but hunky) men into digital superheroes, for instance, Christian Bale into Batman. Sirault, who happens to be quite a hunky Frenchman, got into the machine to try it out. Then he and Furbank asked the friendly Fox engineers if they thought the machine would work for a plant. One of them thought it would and even spent two weeks writing software to prove it, until he was called away to work on another film.

Four years on at the High Resolution Plant Phenomics Centre, it's not hunks who are being digitised in this machine, but potted cotton seedlings. A small army of them trundles along a treadmill into 'PLANT SCAN'.

Here every minuscule aspect of their fuzzy phenotype is being digitised into hard data. A LIDAR camera captures an *exact* 3D representation of their dimensions—something never before possible:

How do you measure the exact dimensions of every leaf and tendril of a plant? At the same time, an infra-red camera creates a heat map of each plant to deliver information on everything from drought and salinity tolerance to photosynthetic rate.[34] Furbank needed a military clearance to buy it because these cameras are the ones that sit on the nose cone of heat-seeking missiles! Last but not least a spectrometer, like the one Furbank imagined while cutting plant stems with a bent steak knife, overlays sugar level readings onto the 3D map of each plant.

Here is Furbank's fantasy on steroids. Want a one-in-ten-thousand wheat plant that not only stores stem sugars, but also grows rapidly under salty or dry conditions? This machine will find it. And to prove it to farmers, who don't trust anything unless they can see it being grown in the field, Furbank and Sirault have designed field robots—like the 'Phenomobile', a half-million-dollar golf buggy decked out with the same cameras that I'd seen in 'PLANT SCAN'. With Sirault at the

A small army of cotton seedlings about to have their characteristics 'digitised'. (Courtesy CSIRO)

wheel this buggy can trundle up and down a thousand plots a day. And then for an aerial view Sirault gets to fly a blimp, held by a string like a kite, that can scan the temperature and greenness (another measure of photosynthesis) of five hundred plots at once. On the drawing board is Google Crop—a project to link the communicative reach of Google Earth to crops growing in the field. The idea is to put sensors in the fields that will send back messages in real time to say how the plants are feeling. Furbank trained as a plant biophysicist and Sirault as a statistical geneticist but these days they mostly work with CSIRO teams from information technology. One team they work closely with is also analysing the information from brain scans to detect early Alzheimer's disease. Whether it's scanning brains or plants or superheroes, it all seems to involve much the same techniques.

In another corner of the Thunderbirds lab, Furbank shows me another set of gadgets. These aren't robotic yet—they need a human operator. One is an airtight chamber that clamps onto a leaf and measures how much carbon dioxide the leaf sucks out. The other is a fluorometer gun that shoots a beam of light at a leaf and measures the red fluorescence emitted from the chlorophyll pigments that comprise the plant's solar panels. Both are sophisticated means of measuring the performance of the plant's photosynthetic engine.

These gadgets, as well as the robots, will play a central part in the global effort to boost rice and wheat yields. My journey through the digital agriculture hothouse was like a tour through the Porsche design headquarters—much of the engineering to create the plant photosynthetic engine of the future will take place here.

**

By 2008, Bob Furbank and his toys were very much in demand. And not just among wheat breeders. The rice breeding community were keen to join forces. With the spike in grain prices and the flattening

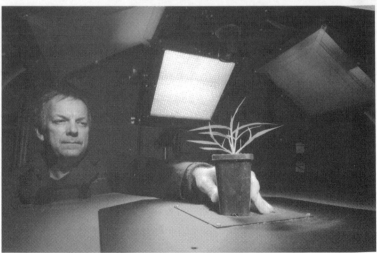

Measuring the performance of the plant photosynthetic engine.
Top: The more CO_2 a plant sucks out of a clamped chamber, the more efficient it is.
Bottom: Robert Furbank measures a plant's fluorescence—the less it fluoresces, the more efficient it is.
(Courtesy CSIRO)

of rice yields, John Sheehy at IRRI was pushing for a project to re-engineer a C_4 engine into rice. But when he invited Furbank to join the global team, the plant engineer was reticent.

With good reason. People had tried before and failed. For instance, Maurice Ku, from Washington State University had transferred some of the genes that were critical to running the corn C_4 engine into rice. At an April 2000 meeting held at IRRI in Los Baños, The Philippines, Ku reported his spectacular result: rice plants with the corn genes delivered 35% more starch! But when Furbank was sent some of this rice to test, he and former boss Hal Hatch (who was dragged out of retirement for the experiment) couldn't find any evidence that the

Rice, by Christine Zavod

enzymes produced by the corn genes were active in these rice plants. Furthermore, when their Japanese collaborators measured the starch level of the re-engineered rice, they found it was not, in fact, more productive than the original variety.[35] Clearly, re-engineering a rice engine was not as simple as transferring a few genes.

Ultimately, Sheehy convinced Furbank of the need for this 'mission impossible'. The spike in rice prices also convinced the Gates Foundation to fund the bold project.[36]

One of the biggest challenges, and the reason why Ku failed, is that you can't just go dumping the components of the C_4 engine into rice. It doesn't have the right infrastructure. It would be like dumping the machinery for a car production plant onto the floor of a furniture factory. Without the conveyer belt, overhead hoists and assembly rooms, the machinery would lie sprawled on the floor, unable to perform. Indeed, the infrastructure of the rice and corn plants is about as similar as a car and a furniture factory.

In a C_4 plant like corn, the two parts of the photosynthesis engine are split across two different cell types.[37] Mesophyll cells use light to produce the C_4 compound, which they ferry to the larger gas-tight bundle sheath cells. Bundle sheath cells unpick the carbon dioxide and concentrate it to turbo-charge Rubisco's sugar output. It's an impressive production line. All the plant's Rubisco is housed in the bundle sheath cells that line the veins, so the sugars they produce funnel straight into the plant's circulation. However, because the engine is split between the mesophyll and bundle sheath cells, the two cell types can never be too far away from each other; specifically, no mesophyll cell (which needs to ferry its C_4 chains) can afford to be more than two cells away from a bundle sheath cell.

The corn leaf has an impressive infrastructure to house its split engine—it's called Kranz anatomy after the German scientist who first observed it. The rice plant is another story. Its bundle sheath and mesophyll cells are about the same size and simply duplicate the same reactions. The bundle sheath cells are not particularly gas-tight and may be as many as eight cells away from the mesophyll cells. One begins to see the difficulty with this retrofitting job. Yet on 62 different occasions, plants evolved a way to do it! That lays down the gauntlet to human ingenuity. Rowan Sage, who studies the evolution of C_4 plants at the University of Toronto, is confident it can be done: 'It's not like a tall mountain; it's more like a bunch of small steps we have to take.'[38]

The consortium is taking a bunch of small steps in many directions at once.[39]

There is the bottom-up approach. Researchers are looking for rice plants that have fewer than eight mesophyll cells between their bundle sheath cells. Deviants would be visible to the eye because their leaf veins (surrounded by bundle sheath cells) would be closer together. At IRRI's centre of operations in Los Baños, technicians have been screening natural and mutant varieties of rice using hand-held digital microscopes—they cost as little as $100 and will not only magnify leaf tissue a hundredfold but also send the images around the globe by email. Over the last year their efforts have identified several rice varieties with *narrower* vein spacing.

Then there is the top-down approach that Furbank refers to as 'stripping down the Ferrari'. In a Ferrari plant like sorghum, the activities of at least 12 genes are tuned to very high levels. And they have the Kranz anatomy: closely spaced veins, big gas-tight bundle sheath cells. But is it all absolutely essential for turbo-charging? To find out, the IRRI researchers have started taking the Ferrari apart. They did it by bombarding a million and a half sorghum plants either with a mutagenic chemical (called EMS) or gamma rays. Then they planted these mutants in a vast sorghum sea and a small army of technicians waded through with their digital microscopes, looking for any runts that perchance also happened to have a *wider* spacing between their leaf veins. They found 17 of them. Now they are being shipped to Black Mountain in Canberra to have their engines rigorously tested at the High Resolution Plant Phenomics centre. Do they splutter if exposed to low carbon dioxide? If so, then they have lost their ability to turbo-charge. What gene was responsible? What changed in their Kranz anatomy? By zeroing in on exactly what is broken, the engineers hope to refine their understanding of what makes the C_4 engine work.

Then there is the lateral approach. Can they catch C_4 evolution in the act? Some plant families are tantalising because they contain both C_3 and C_4 species. Like *Cleome* for instance. *Cleome spinosa* has a C_3 engine, while *Cleome gynandra* has a souped-up C_4. Since their genes are almost identical, it's a matter of 'spotting the difference'. Perhaps they read their genetic script differently? To find out, the researchers are reading the transcriptome—the complete readout of the genome from the two species. The two species may be like two individuals who each possess a largely identical copy of Shakespeare's *Romeo and Juliet*. But while one may be reading the balcony scene to its leaves; the other may be reading the love scene. These different readings will show the researchers which part of the genome script are important for transforming a C_3 engine to a C_4 model.[40]

Finally, another team are rolling up their sleeves and getting on with the retrofit. To emulate the C_4 split-engine structure that eliminates Rubisco from mesophyll cells while dialling up the activity of particular turbo-charger enzymes, the researchers at Cambridge University are employing state-of-the-art genetic engineering. They have made a gene that acts like a guided missile to wipe out Rubisco in the mesophyll cells, and have started installing genes for the C_4 turbocharger into the wreckage.[41] Some of these prototypes have already arrived at Black Mountain. Here the engine redesign will begin in earnest monitored by phenomics robots and precision microscopes. For instance, confocal fluorescence microscopes can identify whether any of the prototypes have split the functions of their mesophyll and bundle sheath cells to resemble those of C_4 plants. And PLANT SCAN will be looking for hotter, faster-growing plants: C_4 plants are warmer and can grow twice as fast as C_3 plants.[42]

With the requirement for coordination back and forth across the globe, this is an engineering effort to rival the one that put a man on the moon. Except it's much cheaper! Rowan Sage at the University of Toronto estimates the total bill at US$100 to $500 million, about the cost of *one* of Canada's intended fleet of F-35 fighter jets.

By 2012, the engineers need a proof of concept to convince the Gates Foundation to renew their funding. Furbank is hoping for a rice plant that is emitting its first wobbly splutters as a C_4 engine. 'If we can't crack it, it's not crackable. Never before has an international consortium had its eyes on the one ball. It may take 25 years but it's our responsibility to take up the challenge', says Furbank.[43]

***

In 2009, wheat got into the act with the International Wheat Yield Consortium driven by CIMMYT in Mexico and funded with US$10 million dollars from the Mexican government. This consortium includes researchers in the UK, the US and Australia.[44] They are not seeking a total C_4 overhaul of the wheat engine. Wheat is tough to re-engineer. For one thing, it has six copies of every gene while rice, corn and most other creatures have two. Five 'back-up copies' make it much harder to disable a particular gene in wheat.[45] Besides, says Furbank, 'We don't want to put all our eggs in one basket. We have a spectrum of projects from long-term ones [like C_4 rice] to short term.'[46] The hope is that the Wheat Yield Consortium will deliver photosynthetically boosted wheat in five years by tinkering with the existing C_3 engine. Remarkably, breeders have never tried for photosynthetically superior plants before, largely because they had no tool to select them. Now they have.

Today in Obregon, a new generation of scientists are wading though fields knee-deep in wheat; the same fields that Norman Borlaug waded through, clipboard and ruler under his arm, meticulously checking each plant for how its leaves curled and whether its tips were white; reading the pattern of flecks and pustules, and many other things. These scientists are not carrying clipboards and rulers, nor are they relying on 'feel'. They are carrying three gadgets to measure the performance of the wheat photosynthetic engine. One looks like a tiny digital camera. It clips onto the plant leaf and reads the fluorescent red signal emanating from it. The smaller the signal, the better the engine: it means the plant is directing that fluorescent energy into photosynthesis. Another clip-on gadget looks like a garlic press. It delivers a whoosh of carbon dioxide to the plant leaf and measures what is left in the press to determine how much carbon dioxide the leaf sucked up. Plants that suck well have good engines. Finally, there's an infrared meter that looks like a toy gun. Aimed at each plant, it measures its temperature. The coldest plants are the fastest photosynthesisers because their pores are open wide to suck in carbon dioxide. The escaping water vapour also cools them.

Soon these promising CIMMYT wheat varieties will be arriving at Black Mountain to join the cotton plants trundling through PLANT SCAN. Those whose engines successfully run the gauntlet will emerge as élite performers. They will be plugged into breeding programs to try and lift the flatlines of wheat yields. Meanwhile, the geneticists will try to identify the genes responsible for the élite photosynthetic performance, information that in turn will feed back on the design process: parts list and dynamometers working hand-in-hand to redesign the engine.

***

I was very fortunate to meet Norman Borlaug when I did. Six months later he passed away. But Borlaug's final wish is being granted. The international community is coming together to resuscitate international agriculture. Fifty years ago Borlaug rallied the world to fight human hunger and fomented a Green Revolution. Today a new community of breeders is rallying under the same dire Malthusian threat. Rebuilding the engine of photosynthesis may seem like an impossible mission. But Borlaug didn't have it much easier. When he first came to Mexico in the mid-1940s to find wheat farmers defeated by rust and hostile to the ideas of scientists, he thought he would never achieve anything. Instead, he fed the world for over 50 years.

On a trip to Obregon for a Wheat Yield Consortium meeting in February 2011, Furbank saw Borlaug's recently erected bronze statue. He had never managed to discuss the rice or wheat projects with him. But if Borlaug's statue could speak, its message to Furbank would no doubt be the same simple one that Borlaug gave to the multitude of breeders he inspired.

'Don't be afraid. Do your best. Don't give up.'[47]

7

Meet Your Ancestor

It's a basic drive to want to find out about your ancestors. Sometimes it can lead to surprises.

In 1883, Franz Schulze got a big surprise when he discovered a one millimetre–long thing crawling up the walls of his seawater aquarium. Most people might have missed the translucent blob. But Schulze, an Austrian sponge expert, was a careful observer and it was clear to him that this speck from the Adriatic Sea was exceedingly odd. Some might have taken it for a marine amoeba. But an amoeba is *one* giant cell that slithers via an unmistakeable streaming motion. This creature crawled on tiny hairs and was composed of *thousands* of cells. Some might have taken it for a flatworm. But flatworms have eyes, a brain, muscles, a gut and a clear axis down the middle. This blob had none of those things. It had a discernible top, a bottom, a middle and a few hairy cells—all in all about five different types of cells. To feed, it crawled on top of a patch of algae or bacteria and dissolved it. It was the simplest animal ever discovered. Schulze named it *Trichoplax adherens*, which translates as 'adhesive hairy plate'.

His colleagues were intrigued. Darwin's *The Origin of Species* was just two decades old and the coda was still resonant: 'from so simple a beginning endless forms most beautiful and most wonderful have been, and are being evolved'.[1]

Could *Trichoplax* be the beginning: the Ur-metazoan?[2]

Schulze's colleague Otto Butschli certainly thought so.[3] But he lost the argument to naturalist and artist Ernst Haeckel. With his mesmerising drawings of animals, plants and embryos, Haeckel had won enormous fame as a populariser of Darwin's theories. In his view, *Trichoplax* could not represent the ancestral metazoan because such a creature would have to have been a swimmer rather than a crawler. (Ironically, decades later *Trichoplax* was found to be a perfectly good swimmer.) The final word went to a German zoologist by the name of Thilo Krumbach. He'd noticed *Trichoplax* in close association with a jellyfish in an aquarium, and decided it must be the infantile stage of that creature. When his boss who'd been working on a textbook, suddenly died, Krumbach's mistaken theory appeared in that textbook.[4] Thus *Trichoplax* lost its allure as the potential ancestor of all animals. Besides which, it was awfully hard to grow, let alone to see.

For 60 years, *Trichoplax* vanished off the scientific radar screen. But in the 1960s, some zoology students at the University of Frankfurt decided to give their boss, Willi Kuhl, an aquarium from the Mediterranean Sea for his birthday. And somebody noticed something cool crawling up the sides. It was tiny and amoeba-like.

Trichoplax was back. And this time, the scientific world was ready for it. Once again, researchers were astounded by the simplicity of the animal. It didn't even have the basic hallmark of animal-kind: a backing for sheets of cells called the basement membrane. *Trichoplax*'s bid for the title of Ur-metazoan was back on the table. And this time the scientists didn't just stare down microscopes in bemusement. They had tools to probe its genetic secrets.

In 2008, this very strange, mostly unheard-of creature joined the select club of animals to have their genomes sequenced. The paper was published in *Nature* magazine[5] and to those few aficionados paying attention, it was a huge shock. *Trichoplax* had a genome not so different from our own. All of the elaborate gene circuitry that goes to laying down a head-to-tail body axis; to making tissue layers

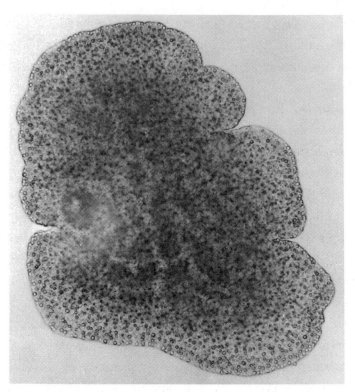

Trichoplax, the simplest animal, is a pancake composed of a few thousand cells. Yet it carries the genes to make complex organs, even a brain. (Photo by Karl J. Marschall, courtesy Vicki B. Pearse)

and basement membranes; to assembling muscles, guts and bones; to hormone signalling; to responses to light, sound and taste—and indeed, most of the circuits for making a human brain—*Trichoplax* had them. What was this pinhead-sized pancake, this slime ancestor, doing with our genes? It was like digging up a Palaeolithic cave and finding crusty silicon chips, transistors and memory cards scattered between the ancient human bones. Our supposed ancestor was all tooled up and ready to go.

But maybe *Trichoplax* isn't really an ancient relative after all. Its extremely simple body might once have been rather complex, putting to good use all those snazzy components that its DNA encodes. Perhaps a more complex *Trichoplax* ancestor found itself a very

comfortable niche somewhere on the rocks, and said, 'Why bother?'
It put most of its genes to sleep and opted for the simple life: the
evolutionary equivalent of a couch potato. Perhaps Krumbach, who
thought it a juvenile jellyfish, was right after all.[6] Couch potatoes are
not uncommon. Take the sea squirt, a tough blobby inhabitant of rock
pools often mistaken for a plant, and that zoologists long considered
to be something like a sea slug—until the late 1860s when a Russian
scientist by the name of Alexander Kowalevsky took a closer look at
their larvae and saw they were nothing like their blobby parents. They
resembled tadpoles with eyes, a brain, a gut and muscles. In other
words, they were chordates (animals with a nerve cord) and far more
closely related to *us* than to sea slugs. But these tadpoles seemed to
have little regard for their high-tech features. After a few days, they
swam headfirst into a rock, stuck there, dissolved their brains and
muscular tails, and became filter-feeding sea squirts.[7] Perhaps the
ancestor of *Trichoplax* had also used its entire genome to make complex structures but at some point of its evolution, stopped bothering.

The argument continues as to whether *Trichoplax* is truly simple
and likely to resemble our ancestor when it first crawled out of the
slime; or whether it's 'secondarily simple', a couch potato version of a
fairly advanced animal.[8]

***

What then of other potential ancestors—like sponges, for instance?
When it comes to a candidate for the ancestor of animals, sponges
have impeccable qualifications. They are the earliest animal-like thing
to leave traces in the fossil record. Rocks formed 600 million years
ago on the ocean floor carry their unmistakable imprimatur. Though
being rooted to the ocean floor makes them seem plant-like, sponges
are indeed animals. They have soft membranes around their cells
rather than the rigid walls of plants or fungi, and like all animals they
rely on capturing their food.

As far as animals go, however, sponges are extremely basic.[9] Like *Trichoplax*, most sponges have no basement membrane underlay to their cells.[10] Nor do their 12 different cell types bind together into tissues like muscle or skin. Rather, *most* of their cells are only loosely bound together and embedded in a jelly matrix with, as far as one can tell, fairly little cooperation between them. Sponges can even be atomised back to individual cells, and they will re-aggregate to form a perfectly happy sponge.[11] Hairy cells called *choanocytes* line the pores of the sponge and like trained oarsmen beat their flagella to flush seawater through. But their impressive oarsmanship is not a result of cooperation; they are just responding to the same external triggers, perhaps sediments in the water or a chemical, like oarsmen responding to the coxswain.

The choanocyte oarsmen are a further testimonial to the sponge's ancient ancestry. In 1841, a French biologist by the name of Dujardin noticed a remarkable resemblance between sponge choanocytes and a free-living single cell known as a *choanoflagellate* (which means collared whip-tailed cell). Sometimes for short periods, these choano-flagellates even got together to form a colony of equals. What a smoking gun! You could see what had happened. A colony got together and decided to stick it out. Perhaps being bigger, they outsized their regular predators.[12] Eventually, larger colonies evolved into the civilisation of diverse cells that we know as an animal. Analysing the genes of choanoflagellates has proven Dujardin correct. Of all the single-celled creatures out there, it is the genes of choanoflagellates that bear the closest resemblance to those of animals.[13]

So the sponge, with its tantalising traces of its single-cell origins, and its 600-million-year fossil record, has a very strong claim to being the closest thing to the Ur-metazoan. Naturally, when its genome was published in 2010 in *Nature*, zoologists looked forward to the revelation of its genetic secrets.[14] Once again, they were in for a shock. Like *Trichoplax* before it, the sponge seems to have most of the genetic hardware it needs to build a complex organism. Like *Trichoplax*, it has most of the genes it needs to make a brain, yet it does not have a single brain cell!

Scientists continue to argue about which creature is closest to being the Ur-metazoan. The sponge and *Trichoplax* are contenders, but so too are comb jellies.[15] As Vicki Pearse, a *Trichoplax* expert at the University of California, told me, 'There's a whole lot of animals at the base of the animal tree. We really don't know the shape of the lowest branches, which is not surprising. It was a long time ago.'[16]

But one thing is for sure. It seems the simplest animals we can find, be they *Trichoplax*, sponges or comb jellies, are all genetically over-featured. They may be blobs, but they have genes for making a sophisticated body—all the complex command and control systems that keep our three-trillion-cell civilisation under check.

You might argue your way out of this conundrum by saying that genetically speaking, these modern representatives of ancient animals don't represent the ancestral condition. The creatures we see in 550 to 600 million-year-old rocks may look like their modern counterparts but we don't know whether their genomes were the same. Maybe these ancient ancestors had really simple genomes? As Mansi Srivastava, the lead author of the *Trichoplax* paper in *Nature* told me, 'A genome is a very dynamic place.'[17]

Nevertheless, there is one inescapable conclusion. If you and some of your distant cousins sport red hair, you can be sure that a common ancestor must have carried the red hair gene. Likewise, if these 'primitive' cousins all possess the genetic toolkit for complex life, then their extremely ancient common ancestor must have bequeathed them these genes.

We are left to conclude that the earliest ancestor of animals possessed the genetic toolkit for complex life. And that's a befuddling conclusion.

Prior to the genomics era, the expectation was that whatever animal first crawled out of the slime would have had a very simple set of genes. Eldon Ball, an American-born geneticist at the Australian

National University in Canberra with a long-standing interest in animal evolution, told me, 'You wouldn't have believed this result. We had the conception then: simple morphology, simple genome. We just don't know how the first animals got so complex.'

Ball confesses to having had thoughts of space junk. Perhaps a meteorite bearing the seeds of complex life?

***

The *Trichoplax* and sponge genomes generated some shocks. But they were really aftershocks. The major shock began several years before when we first started comparing the human genome to that of other animals.

In Darwin's day, our ideas about animals were based on what they looked like—on the outside, on the inside and, importantly, as embryos. Since the late 1990s, we've probed their biology in a new way. Like hackers analysing lines of software code, we delve into their secrets by reading their genomes.

The year 1995 heralded the first full genome. It belonged to the bacterium, *Haemophilus influenzae*.[18] The reading of its 18-million-letter DNA code (comprising 1,700 genes) was a test run for the three-billion-letter human genome. Scientists, mostly in the US and the UK, were racing to develop the fastest and most accurate technology. In 2001, the draft code of the human genome was published. But it was to take another two years before the hackers could finally answer the question: How many genes does it take to make a human being?

Most bets had placed it at around 100,000 to 200,000. Surely we had to be at least 50 to 100 times more complex than *H. influenzae*. The bets were wrong. Year by year, the revised estimate veered dismayingly downwards. The estimates in the 1990s of 100,000 to 200,000 had been revised by 2001 to 60,000; by 2002, to 30,000. By 2003, we hit 23,000. A 2011 check revealed we are now sitting at 22,500 genes.[19]

We are, in fact, only ten times more complex than a bacterium and four times more so than a single-celled yeast. The sad truth is we have about the same number of genes as one of our favourite laboratory models: the roundworm *Caenorhabditis elegans*. Cold comfort that we have a third more than another lab favourite, the fruit fly.

Mercifully, there was a saving grace from this humiliating state of affairs. We humans might have around the same number of genes as flies and worms. But there were certainly *some* genes we had that flies and worms didn't. Flies and worms also had their own unique sets of genes. So evolution was not so much about quantity as *quality*. Different groups of animals had arisen through the invention of new genes. They were dubbed taxonomically restricted genes.[20]

However, even this saving grace did not last long, thanks to the efforts of David Miller.[21]

***

Tropical Townsville on Australia's far north-east coast is not everyone's idea of paradise. Many are defeated by the nine months of stultifying heat and humidity, and the quaint frontier-town ambience is no substitute for the vibrancy of big city life. British-born geneticist David Miller did not expect to last more than a year or so. A wine and food lover, he was rather partial to South Australia, where he'd scored a previous appointment at the Waite Institute, researching bacteria that detoxify chemical spills. He recalls his first outing to the nearby Adelaide Hills, sitting on the veranda of the Hahndorf winery, sipping beautiful wine, gazing out over the vineyards and thinking, 'This is the life. I want to stay here.'

Indeed you would, if you'd grown up in the 1950s in the grungy mining town of Salford on the outskirts of Manchester. It was, he recalls, 'horrific'.

The Townsville job involved analysing the biochemistry of corals; bacteria, with their phenomenal biochemical capabilities, had prepared him for anything. At first, Townsville could not compete with the delights of South Australia. But the call of the Great Barrier Reef would soon drown out the seductive sirens of the Barossa Valley. James Cook University was home to some of the world's greatest experts on coral reefs and the late 1980s was a particularly exciting time to be there. Shortly after Miller arrived, Carden Wallace, John Collins and their teams discovered the mass spawning of coral. In late October, on the fourth and fifth night after a full moon, the clear waters of the reef exploded in milky clouds of coral sperm and eggs. There in the balmy waters a coral orgy was taking place, conducted by the light of the moon.

The wonders of the reef took Miller back to a boyhood vision of paradise. As a teenager in Salford, he'd seen a television documentary on Hans and Lotte Hass, Austria's answer to ocean explorer Jacques Cousteau. Hans was a marine biologist and his wife Lotte was a one-time model who Miller says 'looked like Rita Hayworth'.

But it was not *just* Lotte in her bathing suit and diving tackle that fired the Salford boy's imagination; it was the technicolour splendour of the reef. It seemed to belong to a different planet entirely.

The coral that had entranced Miller years before now piqued his scientific curiosity. They might look like plants but they were primitive animals living in an intimate partnership with single-celled algae called zooxanthellae. Virtually nothing was known about the genes of coral, but Miller felt sure they had a major contribution to make as far as understanding animal evolution.

Meanwhile, in another corner of the animal kingdom, the genes of the humble fruit fly were rewriting the textbook on animal evolution. Miller had been paying close attention.

The traditional thinking was that animals with radically different bodies would have radically different instructions for making them.[22] Compare flies and humans. Flies have an external skeleton made of chitin. They see out of a multifaceted eye and their two-chambered heart sloshes fluid around their body cavity that then seeps back into the heart through one-way valves. Humans have an internal skeleton made of calcium phosphate, a camera eye and a four-chambered heart that circulates blood through an intricate system of blood vessels. Clearly, these animals have come up with totally different solutions to the problems of body support, vision and circulation. But

David Miller's introduction to coral reefs came from Austrian diving celebrities Hans and Lotte Hass. (Courtesy Hans Hass Institute)

things are not always what they seem. In the mid-1980s, fruit fly researchers discovered *homeobox* genes. These genes were consummate draughtsmen. They laid down the fly's body plan on a shapeless ball of embryonic cells, just the way a tailor lays down a pattern on a length of uncut cloth.

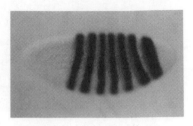

The fushi-tarazu homeobox gene lays down a striped pattern to instruct where to place the segments of the fly embryo. (Courtesy Walter Gehring)

Using this pattern, other genes go to work to 'cut and sew' the cells of the embryo together. Relatives of homeobox genes were soon discovered in nearly every animal. Be they flies, frogs or mice, homeobox genes could be visualised laying down intricate patterns on their embryos. Astoundingly, the homeobox genes seemed to be doing similar jobs in these vastly different animals. For instance, a gene called 'tinman' designs the fly heart and it also has a counterpart in people. An American family with congenital heart defects and fatal arrhythmias turned out to be harbouring a defective gene that was the human equivalent of the fly tinman gene![23] Then there's the 'eyeless' gene, which creates a pattern for the fly eye. Mice and men have a related gene called 'Pax6'. Not only are these genes close relatives, they can do each other's job. When researchers inserted a mouse Pax6 gene into fly embryos, it instructed the formation of a fly eye. The eye formed on whichever part of the embryo was injected—occasionally ending up on the fly's leg![24]

No wonder British geneticist Jonathan Slack dubbed the homeobox gene a 'Rosetta stone'.[25] The ancient Rosetta stone, which carried the same message in readable ancient Greek and inscrutable Egyptian hieroglyphics,[26] was the key to decoding Egyptian hieroglyphics. Likewise, the homeobox genes, which drew body patterns in simple and complex animals, promised to decrypt the mystery of pattern formation in all animals. As a breathless Slack wrote in *Nature* in 1984, 'If so, then we can really anticipate that the problem of pattern formation, one of the deepest mysteries in biology is on the point of final solution.'[27]

Miller was keen to be part of the solution by taking a look at the genes of coral. The timing was good for Australia, too. The country had been left behind in the 11-year race to read the human genome. In catch-up mode, the science-funding agencies agreed to let Miller start peering at the genome of one of Australia's most valuable animals.

It was to take several years for Miller to establish the techniques of extracting and reading coral DNA. He began by reading incomplete snippets of code, enough to give a rough indication of the genes that coral carried.

The genes of coral promised to be very interesting. Compared to flies, worms and humans, it was a far more primitive animal. Fossil members of its extended family (phylum Cnidaria) were found in rocks that were 580 million years old. Their bodies were very simple, comprising an outer and inner sheet with a jelly-like filling between. And though they had neurons, they didn't report to any central brain; they formed a net that relayed environmental signals to the muscles. Miller fully expected that the simple coral would possess a rather simple gene kit.

It didn't. The rough reading revealed an unbelievable inventory of genes. Coral had genes that were supposed to be the exclusive inventions of roundworms. It had genes that were supposed to be the exclusive inventions of fruit flies. But, staggeringly, more than either of these two, the gene set it most closely resembled was that of a human being. Eldon Ball, who collaborates closely with Miller, recalled his own shock at the time. 'It was as if there was a main line going from coral to humans and these two model organisms [fruit flies and roundworms] got off the main line.'[28]

All those 'saving grace genes'—the Taxonomically Restricted Genes of the three species—were there. Coral had it all! The coda of the paper Miller's lab published in 2003 conveys the sense of shock.[29]

Our preliminary survey of the expressed sequences of planula stage *Acropora millepora* appears to turn upside down several preconceived ideas about the evolution of animal genomes. Rather than being simple, the common metazoan ancestor was genetically complex, containing many genes previously considered to be vertebrate innovations. In addition, *Acropora* and human sequences are often surprisingly similar … *Acropora millepora* provides a unique insight into the unexpectedly deep evolutionary origins of at least some vertebrate gene families.

Worse was to come. As more detailed versions of the coral genome were read, they started to reveal really bizarre stuff—things that didn't rightly belong to the animal kingdom at all. 'One set of genes jumped out and poked me in the eye,' Miller told me.[30] For a moment, he must have thought he was back in the bacteria lab in Adelaide. Coral had genes that usually belonged to bacteria. It also had a set of genes

Genes may be lost from the toolkit during evolution

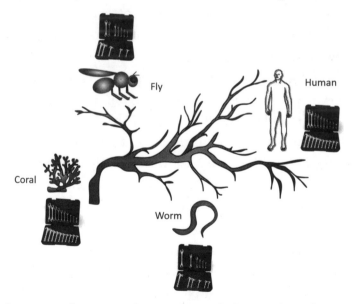

Flies and worms are far more sophisticated animals than coral, yet their genetic toolkits are more poorly stocked. Have they lost these genes during their evolution? Humans have a well-stocked toolkit, surprisingly similar to that of coral. For a list of some of the genes represented in the toolkits, see endnote 29.

that rightly belonged only to plants—the ones that give them their fruity smell.[31]

At first Miller didn't believe the findings. He suspected contamination. Out there on the moonlit reef, who knows what might have contaminated his catch? Lots of things besides coral spawn in unison to overwhelm predators with the sheer mass of numbers. His colleagues were sure that the bonanza of 'coral' genes could not have come *just* from coral. Nipam Patel for instance, a homeobox gene expert visiting from Berkeley University in California, pointed out to Miller that the 'coral' homeobox gene was very much like the vertebrate variety. Surely a fish egg had contaminated the catch.[32] And as for the bacterial and plant genes, well, there was no difficulty in explaining contamination by a few algae and microbes. Miller had made his coral DNA from the larval form to avoid their plant passengers that hop on board at a later stage. But maybe some contamination had already occurred?

Yet, year after year with each new coral spawn, Miller got the same result. Then other reports began to trickle out. Genomes of sea anemones and comb jellies and *Trichoplax* and sponges also turned out to have genes they weren't supposed to have, genes that rightly belonged only to worms, flies or humans, or indeed to the foreign kingdoms of bacteria and plants.[33]

Miller and Eldon Ball began formulating a hypothesis.

***

I heard it six years later. I was attending a genome conference in the lovely seaside town of Lorne, Victoria. Here the tall forests of the Otway ranges meet the surf-tipped waters of Loutit Bay forming a green amphitheatre that resonates with the raucous screeches of white cockatoos. The international invited speakers are invariably enchanted. They are the lucky few selected by the conference organisers as the hottest genome scientists of the year. The hottest of the hot give the opening presentations which I frantically try to get to, somehow always leaving just too little time for the two-and-a-half hour drive from Melbourne.

In 2009, I made it in time to hear the end of the talk from the first keynote speaker who expounded upon the genes of planarians, a type of flatworm. Then it was David Miller's turn.

Miller was a strange speaker. Clad in a colourful surfer's shirt and faded jeans, his ears pierced with gold earrings, his garb was *not* unusual for a genome researcher. It was his manner that was strange. He spoke uneasily, as if labouring under the weight of an idea, as if he half expected to be pelted with eggs.

And rightly so.

A few rows back, I sat listening as he asked his audience to indulge a proposition: 'I'd like to put forward the proposition that in evolution, genes are not gained; they are lost'. It seemed a preposterous proposition, one that as a critical journalist, I might just relegate to the fringe. But he was the second keynote speaker—I was supposed to take him seriously.

I did.

Eighteen months later I escaped Melbourne's chilly spring to visit Miller in balmy Townsville. Palm trees swayed and all was lush and fecund, even extending down below the surface of the tepid emerald waters. There in the depths were stingers and crocodiles, and corals that spawned on a moonlit night. It was their DNA that was disabusing us humans of our sense of superiority, our throne atop the tree of evolution.

***

David Miller's findings had sounded absurd to me when I heard him speak at Lorne: the idea that genes were lost during evolution; that coral represented the ancestral animal with a superset of genes; that subsequent models had pruned away what they didn't need!

But after my day with him in his lab, it dawned on me that his idea was part of a continuum; the latest in a series of upheavals that had come from decoding genomes.

We'd had the Rosetta stone revelation in the 1980s: animals with vastly different body designs still used the same homeobox genes to design them. That showed us that evolution was parsimonious: it didn't reinvent things. Now Miller's findings were showing how incredibly parsimonious the process was. By and large, new species did not evolve by inventing new genes—they played with what was there, and often they ditched some.

As a fruit fly researcher I had been part of the Rosetta stone revolution. I had dug away at the mysteries of my own little fruit fly

Coral has a superset of genes, which not only includes genes once thought to be exclusive to vertebrates, but also to plants and bacteria. (Courtesy David Miller)

homeobox gene called *engrailed*,[34] a gene that the fly used to design its segments, and that vertebrates used to design nervous systems.[35] Glimpsing these revelations spanning two decades now, I was getting a feeling of a bigger picture struggling to emerge. It felt like being part of a murder mystery, where each clue is at first totally bewildering, and then the penny drops.

This was a murder mystery that had been playing out for 150 years. Darwin had been our first Sherlock. In the preface to his magnum opus, *The Origin of Species*,[36] he referred to the 'mystery of mysteries'. Where did new species come from? How had they been created?

In 1859, Darwin offered us evolution. Species changed through a process of descent with modification—not so different to the way dog breeds arise. But, rather than the breeder selecting from the variety of pups in a litter, it was nature doing the job, brutally selecting the fittest to survive.

In the early 1900s, a whopping big new clue was discovered—Mendel's genes.[37] Over the next few decades, a new generation of sleuths like Ron Fisher, Ernst Mayr and J.B.S. Haldane plugged genes into the evolutionary equation to create what was dubbed the Modern Synthesis.[38]

Now the current generation of sleuths have another big new clue to plug in to the mystery: whole genomes. According to evolutionary biologist Eugene Koonin at the National Institutes of Health in Bethesda, Maryland, we are now experiencing the Post-Modern Synthesis.[39]

So today, in the dazzling light of this Post-Modern Synthesis, what is the answer to the 'mystery of mysteries'? How do species change? The answer is bizarre. It seems it doesn't have much to do with *new* genes.

It is of course totally counter-intuitive. Species evolved towards greater complexity: an insect is clearly more complex than a sponge, and a human is clearly more complex than an insect. The natural expectation was that as organisms grew more complex, their genetic toolkit would acquire fundamental new additions. Rather like the way bicycles evolved into cars. Bicycle components comprise wheels, gears, brakes, a frame, a seat and steering. My car has all those things and a whole lot more: cylinders, pistons, a distributor, a battery, a cooling system, a microprocessor, memory cards and stuff I have no idea about. The blueprints for a modern bicycle and a modern car parallel the differences in their complexity.

But the Post-Modern Synthesis is playing havoc with our ideas about evolution. Imagine digging up an ancient blueprint for a bicycle design, and finding that it carried instructions for not only wheels and gears but for microprocessors and memory cards! This would be akin to the shock David Miller experienced.

Miller does not believe the coral spawn was contaminated with the DNA of reef fish, algae or bacteria. He is confident that these are the coral's own genes. Furthermore, it's not just corals that turn out to have these bacterial and plant genes. As more genomes of obscure animals have been read, it turns out these exotic genes show up in a strange spotty pattern among the major subgroupings of animals.[40]

For Miller, the simplest way to interpret these findings is to say that the ancestor of animals had a superset of genes that were lost in a patchy way as other species branched off. So the Ur-metazoan was endowed not only with the toolkit for animal life but with genes from other kingdoms: plants, fungi and bacteria. This is vaguely heretical. But Miller is open to heresy. 'Why not? There was a common ancestor of plants, fungi and animals, and it probably had all sorts of genes to play with. Why couldn't some of those genes have survived into the common ancestor of animals?'

Of course, not everyone agrees with David Miller.

***

There is another explanation as to how coral got its strange genes. Rather than being inherited from a super-endowed ancestor, they were freighted in by bacteria or viruses. This freighting of genes from one species to another is known as horizontal gene transfer. It can occur after a bacterium or virus infects a cell, and some of the bug's genes stay forever. It happens all the time between different species of bacteria, and between bacteria and plants. Indeed, the fact that *Agrobacterium* is keen to insert its genes into plant DNA is the basis of plant genetic engineering.

But gene freighting can also take place between more exotic locations. In an April 2010 study published in *Science*, researchers at the University of Arizona found that a reddish-coloured pea aphid had acquired its pigmentation genes from a fungus.[41] If an aphid can acquire pigmentation genes from a fungus, then what's to stop a coral acquiring genes from a plant or bacterium? Even we are not insulated from gene freighting; 8% of our own genome seems to have originated from virus DNA.

Some researchers suspect horizontal gene transfer may explain how exotic genes turned up in corals. We know that as far as the animal kingdom goes, coral is a bit of a freak: it engages in intimate liaisons with a plant. It's not hard to imagine how these and other trysts may have led to its bonanza of exotic genes. But to Miller, coral's genome is not a freak. Quite the opposite; he believes coral represents the ancestral condition. The fact that most animal groups don't carry these genes reflects a certain randomness to the way genes were inherited: some just got lost along the way.

But Miller admits, 'It remains a very contentious issue. Some of my colleagues don't believe it: they say it would require the ancestor to have had a massive genome.'[42]

If Miller's camp is right, that leaves two glaring questions. How did evolution generate the complex structures of animals if not through inventing new genes? And how is it possible for our slimy ancestor to have acquired such a remarkable toolkit of genes?

***

When it comes to complex structures, our human brain is considered the most complex thing in the universe. So how did animals develop the complexity to move from a sponge to a human brain using the same old genes?

It turns out that researchers have two ready answers.

The first is that genes are like LEGO™ blocks. Using the same blocks, my one-year-old niece can stack them up in a tower while my six-year-old nephew can construct an ornate castle. Something like this appears to have happened in evolution: the operators became more sophisticated. But the coded instructions for these operators don't lie within what was traditionally considered genes; they lie within the so-called 'non-coding DNA'. (See chapter 2.)

The second ready answer is that even if complex animals didn't invent new genes, we know they've been fantastically innovative with what they had.

Once the fundamental genes of the animal toolkit had been invented, evolution seems to have gone modular. Genes have a tendency to accidentally make extra copies of themselves, and once copies exist they can be tinkered with. The standard pliers are there to do the standard jobs. But tinker with the extra copies, and you might evolve a pointy-nosed variety for delicate jobs, a bull-nosed set for bending metal, or a bladed version for stripping wires. In time, your basic set of pliers gets very fancy, yet the designs are derivative rather than new.

Tinkering can produce dramatic results. Take the Hox genes, the draughtsmen that lay down patterns on animal embryos.[43] Corals have one or two. But all the later-evolving animal groups have clusters of eight or more, known as Hox clusters. These genes have been duplicated and tinkered with to spectacular effect: tinkering with Hox genes redesigned the body plans of animals. Some scientists propose it was the evolution of these Hox clusters that drove the Cambrian explosion 540 million years ago—a narrow window of some 10 million years, when the fossil record mushroomed from a few sponges, jellyfish, weird Ediacarans and molluscs to a cornucopia of diverse soft and hard-shelled animals. [44, 45]

Duplication and diversification are probably how the sensory genes of sponges evolved the components of a brain. The circuitry of our brain is made up of well-known components. Your modern genome hacker can recognise their codes as easily as a mechanic recognises what's under the hood of a car: ion channels, G-protein-coupled receptors, cyclic nucleotide coupled receptors, glutamate receptors—many of them embedded in a scaffold referred to as the post-synaptic density (so-called because scientists peering at neurons found this region very densely packed with proteins). A sponge does not have a single neuron, but many of these components are there. They are probably put to work in its juvenile stage, when the sponge larva swims around looking for a new home on the reef using cues from light and chemical signals.

A mother sponge broods her eggs in a pouch and releases them. Within the next 12 hours the gourd-shaped larvae actively find themselves a new home using two types of sensory cells. One set are darkly pigmented with a long hair attached to each cell. They form a pretty ring around the narrow rear end of the larva and they steer the animal using light absorbed by the pigments. As the light gets brighter the hair straightens, sinking the larva. As it gets dimmer the hair goes perpendicular, making the larva more buoyant. This action tends to steer the larva to just the right position on the reef to begin its metamorphosis into a sponge. Another set of 'globular' cells lie dotted along the surface of the larva with their tips poking out into the water. They may be involved in sniffing the environment. Both the pigmented and globular cells are packed with the components that would be used for transmitting nerve signals in other animals.[46] So it looks like a circuit that started out linking a stimulus and response within a single sponge cell, ended up wiring cells together to generate the nervous systems of animals.

What we consider 'brain' genes were probably first used by sponge cells to sense their environment. Over the eons, that basic sensory circuit seems to have duplicated and diversified into a human brain. It's not so different to what engineers achieved with iPhones. With dazzling apps that let you do everything from identify a tune on the radio to translating a French text, iPhones recall Arthur C. Clarke's aphorism, 'Any sufficiently advanced technology is indistinguishable from magic.' Yet my engineer husband assures me that the key components of the iPhone—silicon chips and Boolean logic—were already there in the hulking computing machines of the 1950s.

It's amazing what you can evolve once you have a good basic gadget.

So, how did our ancestor get all those genes? When she first slithered out of the slime, the mother of all animals probably looked something like *Trichoplax*. She was a tiny hairy pancake, with barely more to her than an outside, an inside and some loosely joined cells between. Yet, based on reading the genomes of her closest living relatives, it seems this unprepossessing creature was equipped with an amazing toolkit of genes: virtually everything needed to set animal evolution on a path that would lead to the human brain.

How did she get that way? Space junk has been mentioned—but only in jest.

We do know that as far as the factory floor of life is concerned, our ancestor—the multicellular animal—is a rather recent model. Bacteria and archaea,[47] two major models of single-celled life, were well into production 3.5 billion years ago. Our multicellular ancestor only appeared around 0.6 billion years ago. So it's not hard to imagine that a lot of the basic invention had already taken place.

Let's review what we know about those early production models.

Bacteria and archaea were (and are) great successes. They nourished themselves by fermenting methane and other energy-rich compounds spewing out of volcanos, until three billion years ago when cyanobacteria acquired the genes for photosynthesis. Instead of geothermal energy, they used solar energy to fix carbon dioxide gas into long sugary chains, releasing oxygen as a by-product. As oxygen levels rose, another type of cyanobacteria acquired the genes for respiration, using oxygen to release 15 times more energy from sugars than could be harvested through fermentation.[48]

The kingdoms of bacteria and archaea colonised every corner of the planet: *Pyrolobulus fumaric* basked in the 350 degrees Celsius warmth of undersea thermal vents; Mariana trench barophiles thrived at pressures of 500 atmospheres, *Polaromonas vacuolata* chilled out at up to minus 20 in the polar regions, *Halobacterium salinarium* feasted on salt at the Dead Sea; hydrogen-eating archaea hung out in the geothermal hot springs of Idaho. Others tolerated acids, alkalis, radiation and even arsenic.[49] As the late American palaeontologist Stephen Jay Gould put it, 'For any possible, reasonable or fair criterion, bacteria are—and always have been—the dominant forms of life on Earth.'[50]

Clearly, these 'simple' bugs were assembling an impressive genetic toolkit; one that allowed them to sense and respond to all kinds of environments. Many of the components of that toolkit are just now being revealed as genetic explorer Craig Venter systematically trawls through the world's ocean for novel genes. As Melbourne University microbiologist David Tribe put it, 'Bacteria are not less complex, they are just compact.'[51]

Around two billion years ago, life went into a phase of mergers and acquisitions and a new model of life emerged—the eukaryote. A large cell, perhaps an archaean, mastered the art of movement by evolving a fluid membrane. That also allowed it to engulf prey. For one of its meals, it engulfed one of those bacteria that had the nifty ability to combust fuels using oxygen. Rather than eating it, it entered into a partnership. The bacterium offered its host an inhouse power station. The predator provided the bacterium with food and shelter. In time, the bacterium settled down to become what we now know as the mitochondrion. In a later merger, a cyanobacterium brought in a solar-powered sugar factory, providing this model of eukaryote with the option of making its own food. That bacterium became the chloroplast of plants.[52]

The eukaryotic cell was a vastly improved model of cellular life. Not only was it highly powered, it cordoned off its genome in a membrane-bound nucleus to protect it from further invaders. The eukaryotes came in several models. Some were hunters; some—the ones with chloroplasts—made their own food; some scavenged. Collectively, these single-celled critters were called protists. Eventually they would give rise to three kingdoms of multicellular life: animals, plants and fungi.

But not for another 1.4 billion years!

Many experiments were tried and failed in that long run-up to multicellular life.[53] Much of that period was probably spent honing a command and control system that would serve the needs of a complex society of cells.

But even before our ancestor arrived on the scene, many of the pieces were already in place. Archaea and bacteria had already invented ion channels to sense their environment.[54] Those ion channels would pave the way for the development of nervous systems. Indeed, some 27% of our genetic toolkit had already been evolved by the kingdoms of archaea and bacteria three billion years ago.[55]

Protists had also evolved some fancy new genes. Yeasts, for instance, evolved ways to attract the opposite sex. So-called G-protein-coupled receptors helped them pick the correct mates for themselves. We employ similar receptors for the more prosaic purpose of responding to light, smell and taste.[56] Once a gadget has been invented, it seems evolution just adds different front or back ends, like adding extra bits to your power drill.

But recent findings show that protists were far more tooled up than we ever imagined. Cadherins and integrins are genes that were long considered to be the defining hallmarks of animals. Cadherins hold animal cells together; without them developing tissues fall apart.

Integrins are anchors that hook animal cells to their surrounding matrix and can be hauled in when a cell needs to go roaming. Both proteins sit astride fibres that link cells into a network. Like spiders on a web, these proteins monitor the status of cells in the tissue. When problems occur with the cadherins and integrins, the command and control systems of animal tissues break down, resulting in cancer.[57] But it turns out these genes are not exclusive to animals after all, as the reading of the genome of an ancient protist shows.[58] Choanoflagellates have about 23 cadherin genes, more than most animals. Some of these genes may help hold colonies together when choanoflagellates have their occasional get-togethers. Others seem to be used to harpoon bacterial prey. It's not unlike the way our white cells use their integrins as a grappling hook, as they go about their travels in and out of blood vessels.

So perhaps it's little wonder that by the time the ancestor of animals made its appearance—a colony of hunter cells that found a way to hang together permanently—it was very well tooled-up. So well-tooled, that perhaps it could even afford to prune away some of its components in the course of subsequent evolution.

It's hardly what we expected to find at the base of the evolutionary tree. But that's where reading genomes has taken us.

***

So, meet your ancestor and be humbled: she was not much to look at but on the inside she carried a spectacular genetic dowry thanks to her three-billion-year bacterial, archaeal and protist ancestry. By the time she arrived, most genes had already been invented. The 600 million years since then have just been variations on a theme.

'… from so simple a beginning endless forms most beautiful and most wonderful have been, and are being evolved.'[59] Darwin would have been surprised. Perhaps not such simple beginnings.

Coda

So, a little over a decade on from the first reading, what has been the impact of the Human Genome Project?

There is a sense that a brave new world is upon us; one not unlike that envisaged in the movie *Gattaca*, where to know your genome is to know your destiny. But we are not at *Gattaca* yet. In most cases, researchers can't find the genes for common diseases or traits. So far the genes haven't popped out of studies that scanned 0.03% of the genome—one million sites representing the most common variations or common SNPs. The failure to find the genes has incurred a deluge of criticism for the Human Genome Project. But we're on the cusp of what might be a seismic shift. DNA sequencing costs have plummeted so much that researchers aren't just scanning a tiny fraction of the genome, they are reading genomes in full, and dredging them for genes. Soon they will know whether common and highly heritable diseases like schizophrenia are the result of a few rare genetic mutations or hundreds, maybe thousands, of common ones. If it's a few rare genes, then we will be able to predict schizophrenia from genomes. If it's hundreds or thousands of common interacting genes, then we may never be able to predict schizophrenia. Either way, says Mark McCarthy at Oxford University, 'we should know in about two or three years'. Watch this space and hold onto your hats!

The prospect of living in a *Gattaca* world may provoke anxiety, but the flip side is that finding the genes that predispose us to, or protect us from, disease is a gateway to medical advances. So far, macular degeneration and Crohn's disease are the poster children for the successes of genome scanning. We've discovered that it is genes of the most primitive arm of the immune system—the innate immune system—that predispose people to these diseases. This knowledge is informing new approaches to prevention and treatment. When it comes to the modern plague—HIV—certain individuals have natural defences against the virus. Some individuals have always resisted plagues; otherwise our species would be long extinct. But in the genome generation, we have mined the genes of these individuals to discover their secrets. We've discovered that about one in a hundred Europeans will *never* become infected because they carry two copies of a defective form of the CCR5 gene. The gene acts like an elevator to ferry the virus into the cell, and when it is defective, the virus is defeated at the gates. In a shining example of what genomic mining can deliver, drug developers copied nature's defence by designing a drug called Selzentry™ (also called Maraviroc™) that blocks the CCR5 elevator. Most people, however, do not defeat the virus at the gates. And once infected, if they do not have access to antiretroviral drugs, the virus will destroy their immune systems leading to AIDS and death within ten years. But one in 200 people of various ethnicities will battle the virus successfully for decades. They are the so-called élite controllers. HIV vaccine researchers, regrouping after the failure of three vaccines, are scrutinising the genes of élite controllers to gain new intelligence to battle HIV.

Reading genomes is also paving the way for a brave new world of agriculture by giving us mastery over crop breeding. Ever since we domesticated scrawny weeds to produce wheat and rice around ten thousand years ago, crop breeding has been crucial to sustaining our civilisation. In the 1960s, the Green Revolution gave us rice and wheat varieties that catapulted yields ahead of the population growth rate. However, those breeding techniques, which tinkered with the architecture of a plant, have reached a plateau. Annual yield increases are now less than 1% per year. But our population is heading for nine billion by 2050. Which means we have to raise yields by 1.75% per year to keep up with the 70% increase in demand for food anticipated by the Food and Agriculture Organization of the United Nations. As plant breeders rally to the challenge, one group are going to the heart of the problem by re-engineering the rice and wheat photosynthetic engines for greater output and greater efficiency. Like any engineering effort, they require access to the engine parts (the genomes) and a means to test them (phenomics).

Feeding nine billion people and developing new drugs and vaccines are tangible consequences of reading genomes. But arguably the most powerful impact is the reshaping of scientific ideas. Ideas are powerful. Look at the impact of Darwin's theory of evolution. The reading of genomes is overturning many of our long-held scientific ideas.

In 1957, Francis Crick proclaimed the central dogma: DNA makes RNA makes protein. DNA, the molecule that faithfully carries traits through the generations, was a code for proteins. In 1977, when it became clear that much of our DNA did not code for proteins, those senseless DNA tracts were passed off as junk, a product of foreign invasions or the mischievous tendency of DNA to copy itself. Now we know that 'junk' DNA, some 98.5% of our genome, carries important information. Much of its meaning lies not in protein but in RNA. After decades lying in the shadows, considered no more than a disposable copy of DNA for directing protein recipes, RNA is taking the limelight as we discover its power to control the workings of the genome.

And Lamarck's once laughable theory—that an individual's experience can be passed on through their genes—is back. Researchers have been able to see how an individual's experience can change epigenetic settings—the plastering of methyl groups on DNA that changes the activity of genes. In mice, those changes can be inherited. For instance, what a pregnant mouse eats can change the programming of her babies' genes and those of the next generation. Researchers, still often with a sense of battling the establishment, are testing for similar effects in human beings.

And when it comes to understanding evolution, we have been utterly gobsmacked. First our favourite lab models, fruit flies and roundworms, turned out to have about the same number of genes in their genome as we did. The saving grace was that their genetic toolkit lacked sophistication. But that saving grace vanished when we expanded our reading of animal genomes. The most primitive animals—corals, sponges or *Trichoplax*—not only had similar numbers of genes, but, in the case of coral, toolkits almost as sophisticated as our own and in some cases far better stocked! After two hundred years of guessing and inferring how evolution works, scientists are gazing directly at the DNA codes of animals up and down the evolutionary tree. The data show that ancient and simple animals have very complex gene sets. As far as protein-coding genes, there has not been much innovation in the past six hundred million years. And during evolution, many of these genes have been lost from different species. So far, scientists are struggling to integrate this bombshell into existing evolutionary theory to create the Post-Modern Synthesis.

Perhaps the best answer to the question of what has been the impact of reading genomes is that the age of dogmas is over, as scientists themselves proclaim. 'I promised myself that from now on any bizarre finding in my lab will always be treated with respect, even though it does not make much sense,' said Jean-Michel Claverie, at the University of Mediterranée School of Medicine in Marseilles. 'Given that the change we are going through is so cataclysmic, is beyond what we've ever seen, can any scientific paradigm sustain this scope of shift? Will we keep a unity of thought?' wonders Mark Mehler, at Albert Einstein College of Medicine in New York.

The genome generation is yet to witness the final fallout of the reading of genomes. There are far more questions than answers. But as Aristotle said, 'To know what to ask is already to know half.' My hope is that as we witness the ongoing explosion of new information, this book will empower the reader to *know what to ask*.

Notes

Introduction

1 The Human Genome Project officially began in October 1990 and was slated to take 15 years. In fact, it took 11 years to produce the first draft, which was jointly published in *Nature* and *Science* in February 2001; see <genome.gov/10001763>, viewed 30 June 2011.

2 Mattick, J.S. and Mehler, M.F., 'RNA Editing, DNA Recoding and the Evolution of Human Cognition', 2008, *Trends in Neuroscience*, vol. 31, p. 208.

Chapter 1—The Idea of a Gene

1 Ridley, M., *Genome*, 1999, Fourth Estate, London, p. 13.

2 A report in February 2010 revealed that Chihuahuas and other small dogs originated in the Middle East some 12,000 years ago. The small dog mutations turn out to be in the IGF-1 gene that controls the size of the embryo. All dogs including the small ones appear to have been domesticated from a small population of Middle Eastern wolves at least 20,000 years ago; see <physorg.com/news186176127.html>, viewed 25 May 2011.

3 <history.nih.gov/exhibits/nirenberg/HS1_mendel.htm>, viewed 25 May 2011.

4 Mendel's laws were independently rediscovered and published in 1900 by three scientists. German botanist Carl Correns, Austrian plant breeder Erich von Tschermak-Seysenegg and Dutch botanist Hugo De Vries; see <genome.cshlp.org/content/17/6/669.long>, viewed 25 May 2011; and <dnalc.org/view/16007-Carl-Correns-Hugo-De-Vries-Erich-Von-Tschermak-Seysenegg.html>, viewed 30 June 2011. Prior to that there had certainly been terms to connote the heritable material. In 1868, Darwin coined the term gemmules as the carriers of heritable information in his theory of Pangenesis published in the book *The Variation of Animals and Plants under Domestication*. In 1889, Hugo de Vries referred to 'pangenes' in his book *Intracellular Pangenesis*; see <esp.org/books/devries/pangenesis/facsimile/>, viewed 30 June 2011.

In 1905, English geneticist William Bateson coined the term genetics. Finally, in 1909, Wilhelm Johannsen coined the term gene in a book: Johannsen, W., *Elemente der exakten Erblichkeitslehre*, 1909, Gustav Fischer, Jena. The text is viewable at: <caliban.mpiz-koeln.mpg.de/johannsen/elemente/index.html>, viewed 30 June 2011. The word gene comes from the Greek *genesis* ('birth') or *genos* ('origin').

5 See chapter 1 of Fox Keller, E., *The Century of the Gene*, 2000, Harvard University Press, Cambridge, Mass., p. 2.

6 Rubin, G.M. and Lewis, E.B., 'A Brief History of Drosophila's Contributions to Genome Research', 2000, *Science*, vol. 287, p. 2216.

7 Fox Keller, E., *A Feeling for the Organism*, 1983, W.H. Freeman and Company, New York, p. 2.

8 One band on a polytene chromosome does *not* correspond to one gene. The bands correspond to areas where many genes are either active (light staining) or inactive (dark staining). It is estimated that each band corresponds to 15 to 25 genes; see <ceolas.org/fly/intro.html>, viewed 25 May 2011.

9 <lucasianchair.org/papers/brief.html>, viewed 25 May 2011.

10 Fox Keller, E., *A Feeling for the Organism*, 1983, W.H. Freeman and Company, New York, p. 165.

11 'Delbruck felt frustrated in physics since the great paradoxes had already been resolved. He wanted desperately to make an important discovery and expected that biology might be fertile ground. More specifically, he felt that by finding the ideal simple system for a particular problem and by mounting an all-out assault, a situation might be found in which the known understanding of the natural world would be insufficient to explain the results; new laws of physics would be necessary.' Bergman, K., review of Fischer, E.P. and Lipson, C., *Thinking About Science: Max Delbrück and the Origins of Molecular Biology*, 1988, Norton, New York, in 1988, *Science*, vol. 242, p. 1711.

12 Max Delbruck was inspired by a lecture he heard in 1932, 'On Light and Life' by Danish quantum physicist and philosopher Niels Bohr. Delbruck's writings in turn inspired Erwin Schrödinger to write his 1944 lectures 'What is Life'; see <nobelprize.org/nobel_prizes/medicine/laureates/1969/delbruck/?print=1> and <www.nd.edu/~hps/McKaughan.pdf>, viewed 25 May 2011.

13 As quoted in Fox Keller, E., *A Feeling for the Organism*, 1983, W.H. Freeman and Company, New York, p.160. The full quote included 'Certain large protein molecules … possess the property of multiplying within living organisms, [a process] at once so foreign to chemistry and so fundamental to biology'. I omitted the word 'protein' to avoid confusion. Delbruck seems to have prematurely concluded that it was the protein that was responsible for the bacteriophage's multiplication. In fact, it turned out to be DNA.

14 The origin of the term 'molecular biology' is often credited to English X-ray crystallographer William Astbury, who began exploring the structure of biological molecules, such as the keratin of hair in 1931 and DNA in 1938. Judson, H.F., *The Eighth Day of Creation, Makers of the Revolution in Biology*, 1979, Simon and Schuster, New York, p. 72.

15 Ibid., p. 166.

16 Schrodinger, E., *What is Life?*, 1944, Cambridge University Press, Cambridge.

17 Ibid., chapter 3.

18 Ibid., chapter 4.

19 Judson, H.F., *The Eighth Day of Creation: Makers of the Revolution in Biology*, 1979, Simon and Schuster, New York, p. 47 and Crick, F., *What Mad Pursuit*, 1988, Basic Books, New York, p. 18.

20 Watson, J., *The Double Helix*, 1968, Penguin, London, p. 24.

21 They shared the 1969 Nobel Prize for their work.

22 William Lawrence Bragg (known as Sir Lawrence Bragg), his father William Henry Bragg and Linus Pauling were pioneers of using X-ray crystallography to

crack the structures of salt crystals. Bragg and his father won the 1915 Nobel Prize for Bragg's law, which instructed how to work back from the diffraction pattern of an X-ray beam to the arrangement of the atoms in the crystal; see <nobelprize.org/nobel_prizes/physics/laureates/1915/speedread.html>, viewed 30 June 2011.

23 Fibrous structures like keratin are naturally orderly and lent themselves to producing X-ray diffraction patterns. In 1931, English physicist William Astbury first employed X-ray crystallography to explore the regular structure of keratin in a human hair. Calf thymus DNA is also naturally fibrous, which allowed Astbury to produce the first X-ray images showing a regular structure. See Judson, H.F., *The Eighth Day of Creation: Makers of the Revolution in Biology*, 1979, Simon and Schuster, New York, p. 81.

In 1934, Desmond Bernal, an English crystallographer, first showed that a globular protein like pepsin can also produce a crystalline structure and hence could be studied by X-ray crystallography (ibid., p. 80). The assertion that haemoglobin has 10,000 atoms comes from Sir Lawrence Bragg, as quoted by Horace Judson (ibid., p. 106).

24 <osulibrary.orst.edu/specialcollections/coll/pauling/dna/narrative/page1.html>, viewed 25 May 2011.

25 Judson, H.F., *The Eighth Day of Creation: Makers of the Revolution in Biology*, 1979, Simon and Schuster, New York, p. 100.

26 Ibid.

27 Watson, J., *The Double Helix*, 1968, Penguin, London, p. 35.

28 Ibid., p. 36.

29 Judson, H.F., *The Eighth Day of Creation: Makers of the Revolution in Biology*, 1979, Simon and Schuster, p. 109.

30 Rosalind Franklin, a British crystallographer, remains a controversial figure in the story of the discovery of the structure of DNA for several reasons. Her X-ray photos of DNA, particularly the revelatory 'photo 51', helped Watson and Crick elucidate DNA's structure but they were used without her permission. Some commentators consider that she was also not adequately credited for her role in the discovery, largely as a result of the male-dominated scientific culture of the day. For that reason, she has become an icon of the woman scientist's struggle. Watson, Crick and Wilkins were awarded the Nobel Prize in 1962, after Franklin's death at the age of 38 from ovarian cancer in 1958. A recent play about Franklin by Anna Ziegler is called *Photo 51*; see <the-scientist.com/news/display/57807/>, viewed 25 May 2011.

31 Judson, H.F., *The Eighth Day of Creation: Makers of the Revolution in Biology*, 1979, Simon and Schuster, New York, p. 172.

32 Controversy remains over whether Watson, Crick and Wilkins breached scientific ethics by using Rosalind Franklin's images without her permission. And whether Franklin herself was on track to make the same discovery if only it had not been stolen from her. 'The fact remains that she never made the inductive leap,' wrote Horace Judson on p. 172 of his consummate historiography, *The Eighth Day of Creation: Makers of the Revolution in Biology*, 1979, Simon and Schuster, New York. For a thorough exploration of Franklin's contribution, see Chapter 2.

33 '... to spend 40 years working at the cliff face of a problem—the three dimensional, atomic architecture of the hemoglobin molecule, all 10,000 atoms of it ...'. Judson,

H.F., *The Eighth Day of Creation: Makers of the Revolution in Biology*, 1979, Simon and Schuster, New York, p. 20.

34 To get a sense of Linus Pauling, you can't go past Judson's book; see ibid., p. 70.
    You can also take a look at the presentation speech for his 1954 Nobel Prize in Chemistry; see <nobelprize.org/nobel_prizes/chemistry/laureates/1954/press.html>, viewed 25 May 2011.
    And Philip Ball recently wrote an illuminating retrospective on Pauling's contribution to understanding the chemical bond in terms of quantum physics: Ball, P., 'Pauling's Primer', 2010, *Nature*, vol. 468, p. 1036.

35 <osulibrary.oregonstate.edu/specialcollections/coll/pauling/dna/narrative/page8.html>, viewed 25 May 2011.

36 Judson, H.F., *The Eighth Day of Creation: Makers of the Revolution in Biology*, 1979, Simon and Schuster, New York, p. 156.

37 Ibid., as pointed out by American crystallogapher Jerry Donohue, p. 172.

38 Ibid., p. 173.

39 Ibid., p. 175.

40 'One complete turn would be nearly two yards tall.' Judson, H.F., *The Eighth Day of Creation: Makers of the Revolution in Biology*, 1979, Simon and Schuster, New York, p. 175.

41 Brenner, S., *My Life in Science*, 2001, BioMed Central Limited, London, p. 33.

42 Ibid., p. 29.

43 Fox Keller, E., *A Feeling for the Organism*, 1983, W.H. Freeman and Company, New York, p. 167.

44 Brenner, S., *My Life in Science*, 2001, BioMed Central Limited, London, p. 89.

45 Ibid, p. 89.

46 Marshall Nirenberg passed away on 15 January 2010 while this chapter was being written.

47 Brenner, S., *My Life in Science*, 2001, BioMed Central Limited, London, p. 116.

48 As Crick put it, 'Once information has passed into proteins it cannot get out again'. Fox Keller, E., *A Feeling for the Organism*, 1983, W.H. Freeman and Company, New York, p. 168.

49 The hypothesis hails from French naturalist Jean Baptiste Lamarck, born 65 years before Charles Darwin in 1744.

50 Jacques Monod won the Nobel Prize in 1965 for showing how bacterial genes were regulated; see <nobelprize.org/nobel_prizes/medicine/laureates/1965/monod-bio.html>, viewed 30 June 2011.

51 There were lots of reasons to think the code would be universal. For starters, all living things had the same DNA and protein. By 1961, the code was shown to be interchangeable between bacteria and rabbits. Francis Crick in his Nobel lecture (<nobelprize.org/nobel_prizes/medicine/laureates/1962/crick-lecture.html>, viewed 25 May 2011) refers to the work of Ehrenstein and Lipmann (von Ehrenstein, G. and Lipmann, F., 'Experiments on Haemoglobin Biosynthesis', 1961, *Proceedings of the National Academy of Science*, vol. 47, p. 941), who showed that a haemoglobin messenger RNA from rabbit red blood cell precursors (reticulocytes) worked together with bacterial components (now known to be transfer RNA, the adaptor that matches a triplet of bases to an amino acid) to make haemoglobin protein. By 1977, researchers in Herbert Boyer's lab in San Francisco

(Itakura, K., et al., 'Expression in *Escherichia coli* of a Chemically Synthesized Gene for the Hormone Somatostatin', 1977, *Science*, vol. 198, p. 1056) put the predicted coding sequence of the mammalian hormone somatostatin into *E. coli*. Bacteria and mammals clearly have a common language; the bacterium translated the DNA sequence into the predicted protein, opening the doors to the era of genetic engineering. Boyer founded the world's first recombinant DNA company, Genentech, in 1976. In 1982, the company produced the world's first recombinant DNA-based drug. That was human insulin grown in *E. coli*.

52 Friedmann, H.C., 'From Butyribacterium to *E. coli*: An Essay on Unity in Biochemistry', 2004, *Perspectives in Biology and Medicine*, vol. 47(1), p. 47.

53 <genomenewsnetwork.org/resources/timeline/1957_Crick.php>, viewed 25 May 2011.

54 McClintock's findings in maize suggested that genes could be controlled by DNA sequences outside the gene. In fact, these were mobile bits of DNA that jumped around the genome. If they landed near a particular gene they turned it off, and if they jumped back out again they restored the function of the gene.

55 'Having argued for many years that the gene was merely a figment of geneticist's imaginations, that the genetic unit was the chromosome ...'. Fox Keller, E., *A Feeling for the Organism*, 1983, W.H. Freeman and Company, New York, p. 155.

56 <nobelprize.org/nobel_prizes/medicine/laureates/1975/temin-autobio.html>, viewed 30 June 2011.

57 <genome.cshlp.org/content/17/6/669.long>, viewed 25 May 2011.

58 Mattick, J., 'The Hidden Genetic Program of Complex Organisms', October 2004, *Scientific American*, p. 62; and Mattick, J.S. and Makunin, I.V., 'Non-coding RNA', 2006, *Human Molecular Genetics,* vol. 15 (Review Issue 1), R17.

Chapter 2—'Junk is Telling Us Something'

1 Epigenetics 2009 Australian Scientific Conference, held in Melbourne, 1–4 December 2009.

2 Amaral, P.P., Dinger, M.E, Mercer, T.R., and Mattick, J.S., 'The Eukaryotic Genome as an RNA Machine', 2008, *Science*, vol. 319, p. 1787; Mattick, J.S., 'The Hidden Genetic Program of Complex Organisms', 2004, *Scientific American*, vol. 291, p. 60; and Pollack, A., 'The Promise and Power of RNA', 2008, <nytimes.com/2008/11/11/science/11rna.html>, viewed 30 June 2011.

3 In February 2010, Mattick was awarded the prestigious A$4 million Australia Fellowship. <lifesciencelab.com.au/news/nhmrc-announces-2010-australia-fellowship-winners>, viewed 25 May 2011.

4 Researchers were restricted to studying those genes that made abundant mRNA, for instance, the beta globin gene that produced a subunit of haemoglobin. Other genes studied for the same reason were the silkworm fibroin gene and the chicken egg ovalbumin gene. Robert Williamson, pers. comm., 13 January 2010.

5 To find the gene, researchers used radioactively labelled messenger RNA for beta globin as a beacon. It successfully 'lit up' the gene, showing it to be ten times longer than its messenger RNA.

6 Williamson, R., 'DNA Insertions and Gene Structure', 1977, *Nature*, vol. 270, p. 295.

7 Robert Williamson is now policy secretary for the Australian Academy of Science.

8 Alu is one example of a transposon that seems to have been home-grown. It is closely related to 7SL RNA, an RNA gene whose purpose is to direct the trafficking of proteins; see Ullu, E. and Tschudl, C., 'Alu Sequences Are Processed 7SL RNA Genes', 1984, *Nature*, vol. 312, p. 171; and <wikipedia.org/wiki/7SL_RNA#cite_note-pmid6209580-0>, viewed 25 May 2011.

9 For a review of human transposons and the diseases they cause, see <genomemedicine.com/content/1/10/97>, viewed 25 May 2011.

A recent report links a form of autism called Rett syndrome to jumping genes; see <nature.com/news/2010/101117/full/news.2010.618.html>, viewed 25 May 2011.

10 The largest known genome is that of the amoeba *Polychaos dubium* at 670 billion base pairs. The largest plant genome is that of a type of lily, *Fritillaria asyyrica*, at 130 billion base pairs; and the largest vertebrate genome is that of the marbled lungfish, *Protopterus aethiopicus*, at 130 billion base pairs. *Homo sapiens* has a middling 3 billion base pairs; see <en.wikipedia.org/wiki/Genome>, viewed 30 June 2011.

11 In 1968, biophysicist Roy Britten discovered that the genomes of eukaryotes were riddled with repetitive DNA that did not code for proteins; see Britten, R.J. and Kohne, D.E., 'Repeated Sequences in DNA', 1968, *Science*, vol. 161, p. 529. Britten together with Eric Davidson developed an influential model proposing that the RNA transcribed from repetitive DNA sequences might play the crucial role in regulating the genes of complex organisms and facilitate evolution. Both are now at the California Institute of Technology. 'The existence of repeated sequences in higher organisms led us independently to consider models of gene regulation of the type we describe here. This model depends in part on the general presence of repeated DNA sequences. The model suggests a present-day function for these repeated DNA sequences in addition to their possible evolutionary role as the raw material for creation of novel producer gene sequences.' The authors also proposed a broader definition of a gene: 'A region of the genome with a narrowly definable or elementary function. It need not contain information for specifying the primary structure of a protein.' Britten, R.J. and Davidson, E.H., 'Gene Regulation for Higher Cells: A Theory', 1969, *Science*, vol. 165, p. 349.

In the early 1970s, Francis Crick proposed that RNA molecules would be better equipped than proteins to regulate long stretches of DNA, and wrote a paper titled, 'The Principle of Versatility and Recognition by RNA'; see Brenner, S., *My Life in Science*, 2001, BioMed Central Limited, London, p.147. Sydney Brenner also distinguished between garbage and junk; ibid., p. 148. You can also hear Brenner's own words by going to <webofstories.com/play/52423?o=MS> and selecting '173. The C paradox inspires Francis Crick to ideas about regulation' and '175. Extra DNA is kept like junk, not thrown out like garbage', viewed 30 June 2011.

Physicist Eugene Stanley at Boston University found evidence to support the functionality of junk DNA using a statistical analysis; see Peng, C.-K., 'Long-range Correlations in Nucleotide Sequences', 1992, *Nature*, vol. 356, p. 168.

Malcolm Simons was granted a patent on junk DNA on 25 August 1989 (pers. comm., Malcolm Simons, 10 February 2010). This laid the basis for the biotech company Genetic Technologies.

12 Each person inherits two sets of three HLA genes, HLA-A, HLA-B and HLA-C. They form a kind of uniform that clothes the surface of cells. They are also highly variable genes. To have a good chance of accepting a tissue transplant, a person's HLA genes should be closely matched to that of the tissue donor.

13 Archaea are single-celled life forms often found in extreme environments like undersea volcanoes and until 1977 were considered members of the bacterial kingdom. However, the composition of their membranes and their protein-making machines (ribosomes) were found to be so different from bacteria that Archaea were given their own kingdom. Rather than being closely related to bacteria, many scientists now consider them more closely related to eukaryotes, which include creatures comprised of complex cells like our own. For a good overview and a list of references, see <en.wikipedia.org/wiki/Archaea>, viewed 30 June 2011.

14 Aristotle viewed the role of the brain as being to cool the blood; see <philosophicalmisadventures.com/?p=13>, viewed 30 June 2011; for other metaphors of the brain, see <cwx.prenhall.com/bookbind/pubbooks/morris2/ chapter2/medialib/lecture/brain.html>, viewed 30 June 2011.

15 Karl Popper viewed the role of science as being to try to disprove theories. Thus only those that withstood the assault would survive.

16 Cumberledge, S., et al., 'Characterization of Two RNAs Transcribed From the Cis-Regulatory Region of the abd-A Domain Within the *Drosophila bithorax* Complex', 1990, *Proceedings of the National Academy of Science.*, vol. 87, p. 3259.

17 Mattick, J., 'Introns: Evolution and Function', 1994, *Current Opinion in Genetics & Development*, vol. 4, p. 823.

18 'In any highly competitive system—whether biological or industrial—the speed and efficiencyoforganization,andthesophisticationofresponsetochangingcircumstances are critical determinants of the system's survival and success. We suggest that this is the imperative that results in biological regulatory networks scaling quadratically with system size in order to maintain optimal integration.' Mattick, J.S. and Gagen, M.J., 'Accelerating Networks', 2005, *Science,* vol. 307 p. 856.

19 One reason amoeba, lungfish and lilies have so much more DNA than we do is that they carry multiple copies of their genomes; it's as if they copied their chromosomes but forgot to divide. So while there is more DNA, it is not really more information— just as copying this page ten times would not give you any extra information.

20 See Figure 1 in Mattick, J.S., 'A New Paradigm for Developmental Biology', 2007, *The Journal of Experimental Biology*, vol. 210, p. 1526; Taft, R.J., et al., 'The Relationship Between Non-protein-coding DNA and Eukaryotic Complexity', 2007, *BioEssays*, vol. 29, p. 288; and Mattick, J.S., 'The Hidden Genetic Program of Complex Organisms', 2004, *Scientific American*, vol. 291, p. 60.

21 Tom Cech shared the prize for discovering ribozymes together with Sidney Altman at Yale University. The gene he studied produced RNA for the needs of the ribosome—a so-called ribosomal RNA gene; see <nobelprize.org/nobel_prizes/ chemistry/laureates/1989/cech-lecture.html>, viewed 2 August 2011.

22 They are termed group 1 and group 2 self-splicing introns; see <en.wikipedia.org/ wiki/RNA_splicing>, viewed 30 June 2011.

23 Link, K.H. and Breaker, R.R., 'Engineering Ligand-responsive Gene-control Elements: Lessons Learned From Natural Riboswitches', 2009, *Gene Therapy*, vol. 16, p. 1189.

24 Cech, T., 'The Ribosome Is a Ribozyme', 2000, *Science*, vol. 289, p. 878.

25 In 1956, George Palade observed ribosomes under a microscope and using chemical tests found them to be equal parts protein and RNA. He shared the 1975 Nobel Prize for that work.

26 Even the simplest ribosomes found in bacteria carried 50 proteins entwined with RNA molecules. No-one had ever tried to crystallise anything so huge.

27 Ada Yonath pioneered this fool's errand. People likened it to Icarus's attempt to fly to the sun on artificial wings. X-ray crystallographer James Whisstock, pers. comm., Monash University, 30 March 2010.

28 The 2009 Nobel Prize for chemistry for studies of the structure and function of the ribosome was shared by Ada E. Yonath, Venkatraman Ramakrishnan and Thomas A. Steitz; see <nobelprize.org/nobel_prizes/chemistry/laureates/2009/>, viewed 25 May 2011.

29 <http://en.wikipedia.org/wiki/RNA_world_hypothesis>, viewed 25 May 2011.

30 Davenport, R.J., 'Making Copies in the RNA World', 2001, *Science*, vol. 292, p. 1278.

31 Schmutz is Yiddish for a bit of dirt.

32 <http://discovermagazine.com/2009/oct/03-sea-change-challenging-biology. s-central-dogma/article_view?b_start:int=1&-C=>, viewed 25 May 2011.

33 Matching RNA that could gum-up messenger RNA was also known as 'anti-sense RNA'.

34 Marx, J., 'Interfering With Gene Expression', 2000, *Science*, vol. 288, p. 1370.

35 Fission yeast are an exception.

36 In 1929, American plant pathologist Harold McKinney found that if he infected tobacco plants with a mild strain of tobacco mosaic virus (TMV), they became fully resistant to pathogenic TMV strains. Agronomists took up the technique, using it to protect crops like tobacco, citrus, cucurbits, grapevines and pawpaws against viral infections.

   In animals, immunity is triggered by the virus proteins; the same was assumed to be true for plants. It wasn't. Baulcombe and Waterhouse showed that the agents of immunity were the tiny pieces of 20-letter-long double-stranded RNA that travelled from cell to cell of the entire plant; see <lifescientist.com.au/article/201247/rnai_sound_silence/?pp=1>, viewed 25 May 2011.

37 Ron Plasterk's team at Hubrecht Laboratory in Utrecht, the Netherlands, Andrew Fire at the Carnegie Institution of Washington and Craig Mello of the University of Massachusetts reported that mutations that disabled RNA interference in worms led to abnormal transposon movements. 'Transposons were jumping out all over the place', Plasterk says. 'These experiments tell us that RNAi's function is to protect your genome from transposons.' See Marx, J., 'Interfering With Gene Expression', 2000, *Science*, vol. 288, p. 1370.

38 The US Defence Advanced Research Projects Agency (DARPA) decentralised their information systems to make them nuke-proof, leading to the development of the internet.

39 Esquela-Kerscher, A. and Slack, F.J., 'Oncomirs: MicroRNAs With a Role in Cancer', 2006, *Nature Reviews Cancer*, vol. 6, p. 259.

40 <sciencewatch.com/inter/aut/2008/08-apr/08aprSWCroce/>, viewed 2 August 2011.

41 <www.mirbase.org/cgi-bin/mirna_summary.pl?org=hsa>, viewed 2 August 2011.

42 This is due to a palindromic or mirror-image sequence. The base pair rules makes the RNA strand fold back on itself to form a double-stranded hairpin structure; see <http://en.wikipedia.org/wiki/Small_hairpin_RNA>, viewed 25 May 2011.

43 <nobelprize.org/nobel_prizes/medicine/laureates/2006/adv.html>, viewed 25 May 2011.

44 Esquela-Kerscher, A. and Slack, F.J., 'Oncomirs: MicroRNAs With a Role in Cancer', 2006, *Nature Reviews Cancer*, vol. 6, p. 259.

45 Ibid.

46 Sah, D., 'Therapeutic Potential of RNA Interference for Neurological Disorders', 2006, *Life Sciences*, vol. 79, p. 1773; and <nature.com/news/2010/101123/full/468487a.html>, viewed 30 June 2011. But as this latter news report reveals, there have been some stumbling blocks in delivering effective doses of RNA drugs. Part of the problem is that RNA is notoriously fragile.

47 John Mattick likes to use the metaphor of feed-forward routines when describing how RNA can control the activity of genes. The idea is that as one piece of information is produced, like the RNA transcript, it carries within it a further set of instructions buried in the intron RNA.

48 For a description of feed-forward systems, see <en.wikipedia.org/wiki/Feed-forward#Physiological_feed-forward_system>, viewed 25 May 2011.

49 <newscientist.com/article/mg20126954.100-cost-of-dna-sequencing-falls-to-record-low.html>, viewed 25 May 2011.

50 RNA is first rewritten as DNA using the enzyme reverse transcriptase.

51 John Mattick, pers. comm., 20 January 2010.

52 Asthana, S., et al., 'Widely Distributed Noncoding Purifying Selection in the Human Genome', 2007, *Proceedings of the National Academy of Science USA*, vol. 104, p. 12410; Siepel, A., et al., 'Evolutionarily Conserved Elements in Vertebrate, Insect, Worm, and Yeast Genomes', 2005, *Genome Research* , vol. 15, p. 1034; Bejerano, G., et al., 'Ultraconserved Elements in the Human Genome', 2004, *Science*, vol. 304, p. 1321; Stephen, S., et al., 'Large-scale Appearance of Ultraconserved Elements in Tetrapod Genomes and Slowdown of the Molecular Clock', 2008, *Molecular Biology and Evolution*, vol. 25, p. 402; Guttman, M., et al., 'Chromatin Signature Reveals Over a Thousand Highly Conserved Large Non-coding RNAs in Mammals', 2009, *Nature*, vol. 458, p. 223.

53 There have been several projects aimed at discovering the transcriptome: the earliest was initiated by the biotech firm Affymetrix; see <lifescientist.com.au/article/67558/more_than_meets_eye_affymetrix_chases_transcription/>, viewed 30 June 2011. A consortium initiated by the RIKEN institute took a look at what was happening in mice. They called their project FANTOM for 'functional annotation of mouse'; see <fantom.gsc.riken.jp/4/>, viewed 30 June 2011; a US National Institutes of Health consortium took a look at 1% of the human transcriptome in a project called ENCODE; see ENCyclopedia of DNA Elements, <genome.gov/10005107>, viewed 30 June 2011.

54 Claverie, J.-M., 'Fewer Genes, More Noncoding RNA', 2005, *Science*, vol. 309, p. 1529.

55 Unneberg, P. and Claverie, J.M., 'Tentative Mapping of Transcription-induced Interchromosomal Interaction Using Chimeric EST and mRNA Data', 2007, *PLoS ONE*, vol. 2(2), p. 254. Also, for a recent review, see Gingeras, T.R., 'Implications of Chimaeric Non-co-linear Transcripts', 2009, *Nature*, vol. 461, p. 206.

56 Pearson, H., 'What Is a Gene?', 2006, *Nature*, vol. 441, p. 398.
57 Pennisi, E., 'DNA Study Forces Rethink of What It Means to Be a Gene', 2007, *Science*, vol. 316, p. 1556.
58 ENCODE, an acronym for ENCyclopedia of DNA Elements, was the name of the consortium led by the US National Institutes of Health to thoroughly scan 1% of the genome for all the transcripts produced.
59 Pearson, H., 'What Is a Gene?', 2006, *Nature*, vol. 441, p. 398.
60 Ibid.
61 Ibid; see also a great collection of articles from *The New York Times* musing on the new understanding of a gene: Zimmer, C., 'Now: The Rest of the Genome', 2008, <nytimes.com/2008/11/11/science/11gene.html>; Angier, N., 'Scientists and Philosophers Find That "Gene" Has a Multitude of Meanings', 2008, <nytimes.com/2008/11/11/science/11angi.html?pagewanted=all>; and 'Thoughts on Genes', 2008, <nytimes.com/2008/11/11/science/11genequotes.html>, viewed 30 June 2011.
62 Pers. comm., November 2009.
63 See Table 1 and Table 2 for evidence on the function of non-coding RNA in this 2009 paper by John Mattick, <plosgenetics.org/article/info:doi/10.1371/journal.pgen.1000459>, viewed 25 May 2011.
64 Pers. comm., 29 January 2010.
65 Extracted from Table 1, <plosgenetics.org/article/info:doi/10.1371/journal.pgen.1000459>, viewed 25 May 2011.
66 Check, E., 'Genome Project Turns up Evolutionary Surprises', 2007, *Nature*, vol. 447, p. 760.
67 Pers. comm., June 2011.
68 Amaral, P.P., Dinger, M.E., Mercer, T.R. and Mattick, J.S., 'The Eukaryotic Genome as an RNA Machine', 2008, *Science*, vol. 319, p. 1787.

Chapter 3—Lamarck Returns

1 <ucmp.berkeley.edu/history/thuxley.html>, viewed 27 May 2011.
2 '[When] on board H.M.S. "Beagle", as naturalist, I was much struck with certain facts in the distribution of the organic beings inhabiting South America, and in the geological relations of the present to the past inhabitants of that continent. These facts ... seemed to throw some light on the origin of species—that mystery of mysteries, as it has been called by one of our greatest philosophers.'
Darwin, C., *The Origin of Species*, The Modern Library of New York edition, 1993, p. 18.
3 In his third edition of *The Origin of Species*, Darwin acknowledged 34 thinkers including Aristotle and Lamarck who had also proposed theories as to how species change.

> So what hinders the different parts [of the body] from having this merely accidental relation in nature? As the teeth, for example, grow by necessity, the front ones sharp, adapted for dividing, and the grinders flat, and serviceable for masticating the food, since they were not made for the sake of this, but it was the result of accident. And in like manner as to the other parts in which there appears to exist an adaptation to an end. Wheresoever, therefore, all things together (that is all the parts of one whole) happened like as if they were made for the sake of something, these were

preserved, having been appropriately constituted by an internal spontaneity; and whatsoever things were not thus constituted, perished, and still perish

Aristotle, as quoted by Charles Darwin in a footnote in *The Origin of Species*, The Modern Library of New York edition, 1993, p. 6:

> Lamarck was the first man whose conclusions on the subject excited much attention. This justly-celebrated naturalist first published his views in 1801, and he much enlarged them in 1809 in his *Philosophie Zoologique*, and subsequently, in 1815, in his Introduction to his *Hist. Nat. des Animaux sans Vertébres*. In these works he upholds the doctrine that species, including man, are descended from other species. He first did the eminent service of arousing attention to the probability of all change in the organic, as well as in the inorganic world, being the result of law, and not of miraculous interposition. Lamarck seems to have been chiefly led to his conclusion on the gradual change of species, by the difficulty of distinguishing species and varieties, by the almost perfect gradation of forms in certain organic groups, and by the analogy of domestic productions. With respect to the means of modification, he attributed something to the direct action of the physical conditions of life, something to the crossing of already existing forms, and much to use and disuse, that is, to the effects of habit. To this latter agency he seemed to attribute all the beautiful adaptations in nature;—such as the long neck of the giraffe for browsing on the branches of trees.

4 Winchester, S., *The Map That Changed the World*, 2002, Penguin, London, p. 78.

5 'An alchemical complexifying force drove organisms up a ladder of complexity, and a second environmental force adapted them to local environments through use and disuse of characteristics, differentiating them from other organisms.' Jean Baptiste Lamarck, as quoted in Gould, S.J., *The Structure of Evolutionary Theory*, Belknap Harvard, Harvard, 2002, pp. 187.

6 See <physorg.com/news186176127.html>, viewed 25 May 2011.

7 German scientist August Weismann even did the experiment with white mice. From 1887 to 1889, he cut off their tails for five generations. Nine hundred and one baby mice were born with full-length tails.

8 You can read August Weismann's lucid analysis of 'the supposed transmission of mutilations' in 'Essays Upon Heredity', Chapter VIII, a lecture delivered at the Meeting of the Association of German Naturalists at Cologne, September 1988, p. 431, <esp.org/books/weismann/essays/facsimile/>, viewed 27 May 2011.

9 Huxley, J.S., *Evolution: The Modern Synthesis*, Allen & Unwin, London, 1942.

10 Australian immunologist Ted Steele, for instance, suffered pariah status for decades for proposing that the acquired immunity of an adult could be passed on to their offspring. 'Lamarck's Signature', *Ockham's Razor*, ABC Radio, 1 November 1998, <abc.net.au/rn/science/ockham/stories/s14075.htm>, viewed 27 May 2011. These days his ideas are being treated more reverently and he is making use of new techniques to test them; see Steele, E.J., et al., 'Genesis of Ancestral Haplotypes: RNA Modifications and Reverse Transcription—Mediated Polymorphisms', 2011, *Human Immunology*, vol. 72, p. 283.

11 <time.com/time/health/article/0,8599,1951968,00.html>, viewed 27 May 2011.

12 Gurdon, J.B., et al., 'Sexually Mature Individuals of *Xenopus laevis* From the Transplantation of Single Somatic Nuclei', 1958, *Nature*, vol. 182, p. 64; or watch

'John Gurdon Interviewed by Alan Macfarlane 20th August 2008', <alanmacfarlane. com/ancestors/gurdon.htm>, viewed 27 May 2011.

13 'Office Hours With Randy Jirtle', 22 January 2010, <youtube.com/watch?v=GFK 5xPhkhHM>, viewed 27 May 2011.

14 <nytimes.com/2010/03/16/science/16limb.html?_r=1&ref=science&pagewanted= all>, viewed 27 May 2011.

15 The nucleosome spool is made up of eight interlocking histone proteins, two copies of four different types known as H2a, H2b, H3, H4. Alberts, B., *Molecular Biology of the Cell*, 5th edition, 2007, Garland Science, New York, p. 211.

16 The size of the nucleus is estimated at 4 to 5 microns; ibid., Figure 4.9, p. 201; 30 million nucleosomes per genome; ibid., p. 211.

17 This different packing density is what causes polytene chromosomes; see chapter 1.

18 The main carrier of the chemical tags is the histone H3. The particular combination of chemical tags carried by the histones is called the histone code. Alberts, B., *Molecular Biology of the Cell*, 5th edition, 2007, Garland Science, New York, p. 211.

19 This usually occurs where the letter C is followed by a G, and in the promoter sequence of a gene—the site where gene transcription begins.

20 Dickie, M.M., 'Mutations at the Agouti Locus in the Mouse', 1969, *Journal of Heredity*, vol. 60, p. 20.

21 Gilchrist, S., introduced by R. Williams, 'The Discovery of P Elements', *The Science Show*, 7 February 2009, <abc.net.au/rn/scienceshow/stories/2009/2484333.htm>, viewed 27 May 2011.

22 Lu, D., et al., 'Agouti Protein is an Antagonist of the Melanocyte-stimulating-hormone Receptor', 2002, *Nature*, vol. 371, p. 799. The reason the Agouti sisters differ is that the brunette has been able to silence a rogue 'jumping gene' by methylating it; the blonde has not. The jumping gene plays havoc with the DNA that controls a nearby gene that codes for Agouti signalling protein (ASP). ASP is a kind of circuit breaker that works by blocking receptors. In the hair shaft, ASP blocks the melanocyte stimulating hormone receptor and that stops the release of pigment into the hair. In the brain, ASP blocks melanocortin-4 receptors on neurons and that stops the transmission of a satiety signal. ASP should operate on an intricate on–off cycle, like an Indian smoke signal, so that the hair ends up with the striped pigmented pattern of its South American namesake, the Agouti, and importantly so the animal occasionally stops eating. The jumping gene overrides the smoke signal controls and jams ASP in the on position. That's the situation in the obese blonde sister—no pigment; no controls on eating. The svelte brunette, however, has silenced the jumping gene by methylating its DNA, allowing the ASP to run its smoke signal program. In most mice, the ASP gene does not have a jumping gene as a neighbour. This itinerant DNA moved into the neighbourhood about 40 years ago, as evidenced by the Agouti strain of mice who suddenly started showing highly variable coat colours and body weights. When jumping genes first invade a neighbourhood, they tend to be unruly. In time they are silenced (as are most of the jumping genes that comprise 50% of our genome).

23 Morgan, H.D., et al., 'Epigenetic Inheritance at the Agouti Locus in the Mouse', 1999, *Nature Genetics*, vol. 23, p. 314.

24 Waterland, R.A. and Jirtle, R.L., 'Transposable Elements: Targets for Early Nutritional Effects on Epigenetic Gene Regulation', *Molecular and Cellular Biology*, 2003, vol. 23, p. 5293.

25 Folate, vitamin B12, choline and betaine are components of the chemical pathway that produces S-adenosylmethionine, which in turn provides the reservoir of methyl groups for tacking on to DNA. You can view the chemical pathway at <en.wikipedia.org/wiki/File:Choline_metabolism.png>, viewed 30 June 2011.

26 <sciencedaily.com/releases/2010/11/101117141417.htm>, viewed 27 May 2011.

27 Steegers-Theunissen, R.P., et al., 'Periconceptional Maternal Folic Acid Use of 400 µg per Day Is Related to Increased Methylation of the *IGF2* Gene in the Very Young Child', 2009, *PLoS ONE*, 4(11), e7845, <plosone.org/article/info%3Adoi%2F10.1371%2Fjournal.pone.0007845>, viewed 27 May 2011.

28 Line elements are transposable elements or jumping genes that are active in the human genome and have been linked to Rett's syndrome. On 9 March 2010, I put the following question to Fred (Rusty) Gage by email: '… I wondered if the ability of folic acid to silence genes, particularly transposable elements in agouti mice, might also explain the protective effect of folic acid in neural tube closure defects. Could neural tube defects be a result of excessive LINE activity during neurogenesis?'

     His email response on 13 March 2010 was, 'You have proposed an interesting hypothesis that I had not considered. It is testable, at least to see if Folate decreases normal Line activity. Note sure if I would recommend making a formal hypothesis at this point, but then again I am conservative.'

29 Cropley, J., et al., 'Germ-line Epigenetic Modification of the Murine Avy Allele by Nutritional supplementation', 2006, *Proceedings of the National Academy of Science USA*, vol. 103, p. 17308. (According to Cath Suter, Randy Jirtle's group didn't replicate this effect but they also didn't do the experiment the same way. Suter started with a selected population that are mostly lean and brunette to start with. Jirtle used a mixed population.)

30 It's called DNA methyltransferase 3, or Dnmt3, and it is the gene charged with placing methyl groups on newly made DNA.

31 Pers. comm., Ryszard Maleszka, 23 February 2011.

32 Ryszard Maleszka's group at the Australian National University in Canberra turned off the gene that methylates DNA (DNA methyltransferase 3 or Dnmt3) by feeding grubs interfering RNA (RNAi) targeted to that gene. The result was that they produced queens. See Kucharski, R., et al., 'Nutritional Control of Reproductive Status in Honeybees via DNA Methylation', 2008, *Science*, vol. 319, p. 1827.

33 Ryszard Maleszka told me that phenylbutyrate acts as a histone deacetylase (HDAC) inhibitor. Inhibiting the deacetylation of histones is known to activate genes. Pers. comm., 23 February 2011.

34 Peter Gluckman and Mark Hanson from Southampton University have penned their theory of how the foetus is programmed to match its future life in a popular book titled *Mismatch: The Lifestyle Disease Time Bomb*, 2006, Oxford University Press, Oxford. Their reviews in the scientific literature include Gluckman, P.D. and Hanson, M.A., 'Living With the Past: Evolution, Development, and Patterns of Disease', 2004, *Science*, vol. 305, p. 1733; and Gluckman, P.D., et al., 'Towards a New Developmental Synthesis: Adaptive Developmental Plasticity and Human Disease', 2009, *The Lancet*, vol. 373, p. 1654.

35 According to L.H. Lumey, this piece of history is muddied. The Dutch government in exile had also ordered a rail strike. He says 'the question remains as to why the strike was not called off by the Dutch govt in late 1944 when it was clear it

no longer served a military purpose and potentially greatly endangered the food supplies'. Pers. comm., 20 April 2011.

36 By 26 November 1944, official rations, which eventually consisted of little more than bread and potatoes, had fallen below 1000 kcal per day, and by April 1945, they were as low as 500 kcal per day. Widespread starvation was seen especially in the cities of the western Netherlands. Food supplies were restored immediately after liberation on 5 May 1945.

Lumey, L.H., et al., 'Cohort Profile: The Dutch Hunger Winter Families Study', 2007, *International Journal of Epidemiology*, vol. 36, p. 1196; <ije.oxfordjournals. org/cgi/content/full/36/6/1196?ijkey=fb35724534489a73d489954b7ffec1396652 99a1&keytype2=tf_ipsecsha>, viewed 27 May 2011.

37 They were 65% more likely to be obese. Ravelli, G.P., et al., 'Obesity in Young Men After Famine Exposure *in utero* and Early Infancy', *New England Journal of Medicine*, 1976, vol. 295, p. 349.

38 Lumey, L.H., et al., 'Cohort Profile: The Dutch Hunger Winter Families Study', *International Journal of Epidemiology*, 2007, vol. 36, p. 1196, <ije.oxfordjournals. org/cgi/content/full/36/6/1196?ijkey=fb35724534489a73d489954b7ffcc1396652 99a1&keytype2=tf_ipsecsha>, viewed 27 May 2011.

39 Painter, R.C., et al., 'Early Onset of Coronary Artery Disease After Prenatal Exposure to the Dutch Famine', *American Journal of Clinical Nutrition*, 2006, vol. 84(2), p. 322.

40 For a recent review of the famine data, see Lumey, L.H., et al., 'Prenatal Famine and Adult Health', 2011, *Annual Reviews of Public Health*, vol. 32, p. 24.

41 Barker, D.J.P. and Osmond, C., 'Low Birth Weight and Hypertension', 1988, *British Medical Journal*, vol. 297, p. 134; Barker, D.J.P., et al., 'Weight in Infancy and Death From Ischaemic Heart Disease', 1989, *The Lancet*, vol. 2(8663), p. 577.

42 Barker, D.J.P., 'Fetal Origins of Coronary Heart Disease', 1995, *British Medical Journal*, vol. 311, p. 171. Also, Barker, D.J.P., 'Fetal Nutrition and Cardiovascular Disease in Later Life', 1997, *British Medical Bulletin*, vol. 53, p. 96.

43 Pers. comm., 17 April 2011.

44 The PPAR-γ and glucocorticoid receptor genes showed altered methylation in the offspring of undernourished rats as described in this review, Gluckman, P.D., et al., 'Metabolic Plasticity During Mammalian Development is Directionally Dependent on Early Nutritional Status', 2007, *Proceedings of the National Academy of Science, USA*, vol. 104, p. 12796.

45 A report on the paper by the author is available at <news.sciencemag.org/ sciencenow/2011/04/why-skinny-moms-sometimes-produc.html?ref=ra>, viewed 27 May 2011. See also this reference for the paper: Godfrey, K.M., et al., 'Epigenetic Gene Promoter Methylation at Birth is Associated With Child's Later Adiposity', 2011, *Diabetes*, vol. 60, p. 1528.

46 By comparison, alterations of the genetic code (polymorphisms) at genes like FTO or MC4R explain less than 2% of the variation of body fat (from Genome Wide Association studies where the gene variations are compared in populations of fat versus thin people). See Bogardus, C., editorial, 'Missing Heritability and GWAS Utility', 2009, *Obesity*, vol. 17, p. 209; <nature.com/oby/journal/v17/n2/full/ oby2008613a.html#bib10>, viewed 27 May 2011.

47 Keith Godfrey suspects that the mostly likely cause of the different methylation patterns in the children's retinoid X receptor-α gene is the mothers' diets. His reason

for believing that these changes are not driven by the children's own genes is that they failed to find differences in the DNA sequences of the retinoid X receptor-α gene between the children. And the link between cord gene methylation, childhood obesity and mother's diet was independent of the mother's own obesity, suggesting it was not simply an inherited trait. However, Godfrey acknowledged that they had not ruled out the possibility that other genes in the babies' genomes were driving the differences in methylation. Pers. comm., 21 April 2011.

48 According to Peter Gluckman, pers. comm., March 2011.

49 Heijmans, B.T., et al., 'Persistent Epigenetic Differences Associated With Prenatal Exposure to Famine in Humans', 2008, *Proceedings of the National Academy of Science USA*, vol. 105, p. 17046.

50 When the maternal copy of IGF2 is not silenced, babies are born oversized and with a higher risk of childhood cancer, a disease known as Beckwith-Wiedemann syndrome; see Robertson, K.D., 'DNA Methylation and Human Disease', *Nature Reviews Genetics*, 2005, vol. 6, p. 597. And in adults, the IGF2 gene has also been implicated in diabetes and heart disease.

51 Rampersaud, G.C., et al., 'Genomic DNA Methylation Decreases in Response to Moderate Folate Depletion in Elderly Women', 2000, *American Journal of Clinical Nutrition*, vol. 72, p. 998.

52 Based on an extrapolation from a population in Beijing, 10% have frank diabetes. Another 15% have high blood sugar levels, a precursor to type 2 diabetes. See Normile, D., 'A Sense of Crisis as China Confronts Ailments of Affluence', 2010, *Science*, vol. 328, p. 422; Yang, W., 'Prevalence of Diabetes Among Men and Women in China', 2010, *New England Journal of Medicine*, vol. 362, p. 1090.

53 Pers. comm., December 2009.

54 Germ cells are an animal's most precious cells because they carry the genes for the next generation. In the early stages of embryonic development they are quarantined from the rest of the embryo in a structure that will give rise to either a testis or an ovary. Eventually, germ cells give rise to sperm or eggs, depending on the instructions they get from the testis or ovary.

55 From Darwin's 1868 book, *The Variation of Animals and Plants Under Domestication*, as described in Bowler, P., *Evolution: The History of an Idea*, 2009, University of California Press, 3rd edition, p. 254.

56 George Bernard Shaw wrote the play *Back to Methuselah* in 1921 as a celebration of Lamarckism and a diatribe against Darwinism. An excerpt from the preface (Penguin, 1990 edition, p. 31) describes Darwinism as a 'chapter of accidents' and continues:

> As such, it seems simple, because you do not at first realize all that it involves. But when its whole significance dawns on you, your heart sinks into a heap of sand within you. There is a hideous fatalism about it, a ghastly and damnable reduction of beauty and intelligence, of strength and purpose of honour and aspiration, to such causally picturesque changes as an avalanche may make in a mountain landscape … the explorer who opened up this gulf of despair, far from being stoned or crucified as the destroyer of honour of the race and the purpose of the world, was hailed as Deliverer, Saviour, Prophet, Redeemer, Enlightener, Rescuer, Hope Giver, and Epoch Maker; while poor Lamarck was swept aside as a crude and exploded guesser hardly worthy to be named as his erroneous forerunner.

Arthur Koestler's homage to Lamarckism was a book published in 1971 titled *The Case of the Midwife Toad*. Richard Dawkins discusses these authors in *The Blind Watchmaker*, 1986, Penguin, 1991 edition, p. 291.

57 Pennisi, E., 'The Case of the Midwife Toad: Fraud or Epigenetics?', *Science*, 2009, vol. 325, p. 1194.

58 'The Soviets were … opposed to the determinism of Mendelian genetics, which they took to imply that social reform would be ineffective in improving human character.' Bowler, P., *Evolution: The History of an Idea*, University of California Press, 2009, p. 246.

59 Vavilov's insight was to realise that domesticated plants originated from those parts of the world that carried the greatest species diversity. For example, according to his theory, the 'centre of origin' of wheat was the Middle East and Ethiopia.

60 Weissman, I., 'The Ghost of Lysenko', 17 June 2005, *Financial Times*, <ft.com/cms/s/1/dc78b722-df20-11d9-84f8-00000e2511c8.html#axzz1E06X7jMn>, viewed 30 June 2011.

61 Bacteria have a lot of methylated DNA; bacteriophage little. This underpins the strategy for how bacteria defend themselves against their parasites. Bacteria produce DNA-cleaving enzymes, known as restriction enzymes, that attack DNA without methyl groups. It was the use of restriction enzymes to cut and paste together DNA of any type that ushered in the genetic engineering revolution.

62 Razin, A., et al., 'Variations in DNA Methylation During Mouse Cell Differentiation *in vivo* and *in vitro*', 1984, *Proceedings of the National Academy of Science USA*, vol. 81, p. 2275. The last line of the abstract reads, 'Although these changes in DNA methylation seem to be an integral part of the differentiation process, its relation to specific gene expression is still unclear'.

63 <hopkinscoloncancercenter.org/CMS/CMS_Page.aspx? CurrentUDV=59&CMS_Page_ID=0B34E9BE-5DE6-4CB4-B387-4158CC924084>, viewed 30 June 2011.

64 Robert Weinberg discovered that tumours carry a hyperactive accelerator gene called Ras; see <web.wi.mit.edu/weinberg/pub/>, viewed 30 June 2011. During Weinberg's visit to the Hebrew University of Jerusalem in the mid-1980s, Szyf queried him as to whether methylation might be involved in cancer. As Szyf put it, 'There was no evidence but I thought, why not?'. The question, it seems, was not well received. Interview, Melbourne, December 2009.

65 Szyf, M., et al., 'cis Modification of the Steroid 21-hydroxylase Gene Prevents Its Expression in the Y1 Mouse Adrenocortical Tumor Cell Line', 1990, *Molecular Endocrinology*, vol. 4, p. 1144. Note that Feinberg and Vogelstein had shown that cancer cells were hypomethylated in 1983: Feinberg, A. and Vogelstein, B., 'Hypomethylation Distinguishes Genes of Some Human Cancers From Their Normal Counterparts', 1983, *Nature*, vol. 301, p. 89.

66 Jones, P.A. and Baylin, S.B., 'The Fundamental Role of Epigenetic Events in Cancer', 2002, *Nature Reviews Genetics*, vol. 3(6), p. 415.

67 <methylgene.com>, viewed 27 May 2011.

68 Shabason, J.E., et al., 'HDAC Inhibitors in Cancer Care', 2010, *Oncology*, vol. 24(2), p. 180.

69 Meaney, M.J., 'Maternal Care, Gene Expression, and the Transmission of Individual Differences in Stress Reactivity Across Generations', *Annual Review of Neuroscience*, vol. 24, p. 1161.

70 Weaver, I., et al., 'Maternal Care Effects on the Hippocampal Transcriptome and Anxiety-mediated Behaviors in the Offspring That Are Reversible in Adulthood', 2006, *Proceedings of the National Academy of Science*, <pnas.org/content/103/9/3480.long>, viewed 27 May 2011.

71 McGowan, P.O., et al., 'Epigenetic Regulation of the Glucocorticoid Receptor in Human Brain Associates With Childhood Abuse', 2009, *Nature Neuroscience*, vol. 12, p. 342; and interview with Moshe Szyf at an epigenetics conference, Melbourne, December 2009.

72 The research that links behavioural modification to epigenetic changes is also highly controversial. For a non-technical discussion, see Buchen, L., 'In their Nurture', 2010, *Nature*, vol. 467, p. 146.

73 Bygren, L.O., et al., 'Longevity Determined by Ancestors' Overnutrition During Their Slow Growth Period', 2001, *Acta Biotheoretica*, vol. 49, p. 53; and Kaati, G., et al., 'Cardiovascular and Diabetes Mortality Determined by Nutrition During Parents' and Grandparents' Slow Growth Period', 2002, *European Journal of Human Genetics*, vol. 10, p. 682.

74 A subsequent study by these authors in a modern British population also picked out a correlation between smoking by prepubescent boys and obesity in their sons and grandsons. Pembrey, M., et al., 'Sex-specific, Sperm-mediated Transgenerational Responses in Humans', 2005, *European Journal of Human Genetics*, vol. 14, p. 159; and commentary, Whitelaw, E., 'Epigenetics: Sins of the Fathers, and Their Fathers', 2006, *European Journal of Human Genetics*, vol. 14, p. 131.

75 Agrawal, A.A., et al., 'Transgenerational Induction of Defences in Animals and Plants', 1999, *Nature*, vol. 401, p. 60.

76 Robaire, B., 'Is it my grandparents' fault?', 2008, *Nature Medicine*, vol. 14, p. 1186.

77 Dolinoy, D.C., et al., 'Maternal Nutrient Supplementation Counteracts Bisphenol A-induced DNA Hypomethylation in Early Development', 2007, *Proceedings of the National Academy of Sciences USA*, vol. 104, p.13056.

78 'Office Hours With Randy Jirtle', 22 January 2010, <youtube.com/watch?v=GFK5xPhkhHM>, viewed 27 May 2011.

79 Junko, A., et al., 'Transgenerational Rescue of a Genetic Defect in Long-Term Potentiation and Memory Formation by Juvenile Enrichment', 2009, *The Journal of Neuroscience*, vol. 29, p. 1496.

80 The MLH-1 gene repairs damaged DNA and so guards against potential cancer-causing mutations.

81 Suter's first paper showed that the MLH-1 epigenetic change was present in a man with colorectal cancer and his sperm; see Suter, C.M., et al., 'Germline Epimutation of *MLH1* in Individuals With Multiple Cancers', 2004, *Nature Genetics*, vol. 36, p. 497. A subsequent paper showed that the MLH-1 defect was indeed passed on to the next generation—a mother passed on the defect to her son, but his sperm did not show the defect; see Hitchins, M.P., et al., 'Inheritance of a Cancer-Associated MLH1 Germ-Line Epimutation', 2007, *New England Journal of Medicine*, vol. 356, p. 697.

82 Chong, S., et al., 'Heritable Germline Epimutation is Not the Same as Transgenerational Epigenetic Inheritance', 2007, *Nature Genetics*, vol. 39, p. 574; Horsthemke, B., 'Heritable Germline Epimutations in Humans', 2007, *Nature Genetics*, vol. 39, p. 573; Suter, C.M. and Martin, D.I.K., 'Reply to "Heritable

Germline Epimutation is Not the Same as Transgenerational Epigenetic Inheritance"', *Nature Genetics*, 2007, vol. 39, p. 575.

83 Dawkins, R., *The Blind Watchmaker*, 1986, Penguin, New York, p. 293.

84 It is named after August Weismann who famously amputated the tails of mice to show that the experience of the adult was not inherited by the offspring; see <en. wikipedia.org/wiki/Weismann_barrier>, viewed 30 June 2011.

85 For more on the role of piRNA in flies, see 'Not "Junk DNA" After All: Tiny RNAs Play Big Role Controlling Genes', 2007, *ScienceDaily*, <sciencedaily.com/ releases/2007/10/071025112059.htm>, viewed 30 July 2011; and <sciencedaily. com/releases/2008/11/081127145145.htm>, viewed 27 May 2011. And for a review of the role of piRNA in all animals, see Ghildiyal, M. and Zamore, P.D., 'Small Silencing RNAs: An Expanding Universe', 2009, *Nature Reviews Genetics*, vol. 10, p. 94; and Aravin, A., et al., 'The PIWI-piRNA Pathway Provides an Adaptive Defense in the Transposon Arms Race', 2007, *Science*, vol. 318, p. 761; and Moazed, D., 'Small RNAs in Transcriptional Gene Silencing and Genome Defence', 2009, *Nature*, vol. 457, p. 413.

86 The fly's DNA packing crew are called PIWI proteins. PIWI stands for 'p-element induced wimpy (testicles)'. It is a protein that was discovered in mutant fruit flies, who had shrivelled testicles. The defect was a result of overactive jumping genes known as p-elements. The normal gene produced a PIWI protein that silenced the p-elements. Hence the gene/protein was named by describing what happened to the fly when the gene was defective: wimpy testicles. As recounted to me by Haifan Lin of Yale University, discoverer of the PIWI system, pers. comm., 12 March 2010.

87 Suter found that the MLH-1 gene was shut down in colon cancer patients and their offspring. She speculated on how an errant piRNA could shut down a gene over generations. The errant RNA could amplify itself through what is known as the 'ping-pong mechanism', so that enough is carried over to shut down the gene in the next generation. 'It's just a theory, but all you have to do is shut down two genes (the maternal and paternal copy).' Pers. comm., 25 April 2011.

88 Mello, C., 'Return to the RNAi World: Rethinking Gene Expression and Evolution', Nobel Lecture, 2006; Ueki, S. and Citovsky, C., 'RNA Commutes to Work: Regulation of Plant Gene Expression by Systemically Transported RNA Molecules', 2001, *BioEssays*, vol. 23, p. 1087.

89 If mice carry a mutant version of a gene called KIT, they develop spotty tails. Just one copy of the mutant gene will have this effect, even if mice possess a good copy inherited from the other parent. So far this just sounds like a case of a dominant gene. But the strange thing is that if spotty-tailed mice are then mated with normal mice, all their offspring emerge spotty-tailed—even the ones that have two perfectly good copies of the KIT gene. This is a blatant transgression of Mendel's laws. It's as if a blue-eyed mother mated with a brown-eyed father, and their offspring had blue eyes forever after—even those who ended up with brown-eyed genes. Somehow the mutant KIT gene squelches the output from the good copy of the gene. Plant geneticists had seen a similar phenomenon in maize over 50 years ago, which they termed 'paramutation'. Paramutations may eventually flip back to their normal state, so they are less stable than true DNA-based mutations. See Brink, R.A., 'A Genetic Change Associated With the R Locus in Maize Which is Directed and Potentially Reversible', 1956, *Genetics*, vol. 41, p. 872; and Chandler, V. and

Alleman, M., 'Paramutation: Epigenetic Instructions Passed Across Generations', 2008, *Genetics*, vol. 178, p.1839.

  In the case of these spotty mice, Minoo Rassoulzadegan at the University of Nice, France, identified what this 'squelching' factor might be. The sperm (and other cells) of these spotty mice carried some unusual RNA. When she injected this RNA into the fertilised eggs of normal mice, these mice also acquired spotty tails! Her extraordinary finding was published in *Nature* with the demure title, Rassoulzadegan, M., 'RNA-mediated Non-Mendelian Inheritance of an Epigenetic Change in the Mouse', *Nature*, 2006, vol. 41, p. 469. A BBC story on the finding cut to the chase with 'Spotty mice flout genetics laws': <news.bbc.co.uk/2/hi/science/nature/5011826.stm>, viewed 5 August 2011.

90 It was mighty startling in 1987 when researchers first discovered cells of the intestine editing RNA. The end result was that the script for the protein apolipoprotein B (the protein that determines our risk of heart disease) was chopped to about half its usual length. The editor enzyme, APOBEC 3G, works by changing the letter C to U via a chemical reaction called cytidine deamination. See Powell, L.M., 'A Novel Form of Tissue-Specific RNA Processing Produces Apolipoprotein-B48 in Intestine', 1987, *Cell*, vol. 50, p. 831; Chen, S.H., 'Apolipoprotein B-48 is the Product of a Messenger RNA With an Organ-Specific In-frame Stop Codon', 1987, *Science*, vol. 238, p. 363.

91 The editing of serotonin receptor RNA is carried out by an editor enzyme called ADAR. It changes an adenosine residue to an inosine, which is read as if it were a guanosine. Nishikura, K., et al., 'Functions and Regulation of RNA Editing by ADAR Deaminases', *Annual Review of Biochemsitry*, 2010, vol. 79, p. 321. For papers on the editing of serotonin receptors, see Englander, M.T., et al., 'How Stress and Fluoxetine Modulate Serotonin 2C Receptor Pre-mRNA Editing', 2005, *Journal of Neuroscience*, vol. 25, p. 648; Schmauss, C., et al., 'The Roles of Phospholipase C Activation and Alternative ADAR1 and ADAR2 Pre-mRNA Splicing in Modulating Serotonin 2C-Receptor Editing in vivo', 2010, *RNA*, vol. 16, p. 1779.

  Not all strains of mice respond to stress by editing their serotonin receptor ($5HT2_c$). The strain of mice that does respond this way is called BALB/c. They have chronically low levels of serotonin in their forebrain and increase their level of serotonin in response to stresses such as being tipped into a water maze. That also increases editing of $5HT2_c$ receptor to a less active form. The senior author of these studies, Claudia Schmauss at Columbia University in New York, believes editing is an adaptive response to maintain the status quo. That may seem counterintuitive since serotonin is perceived as a calming hormone and you'd think stressed mice would want to calm themselves. Schmauss warns that this is a murky area: 'What calms the mouse down? It's not easy to answer.' Serotonin may increase stress levels, for instance, there's evidence that serotonin signalling via the $5HT2_c$ receptor in a brain structure called the amygdala may increase anxiety. See Nichols, D.E. and Nichols, C.D., 'Serotonin Receptors', 2008, *Chemical Reviews*, vol. 108, p.1614. And a new type of antidepressant drug called tianeptine actually reduces serotonin levels. Schmauss suspects that in the stressed mouse and in chronically depressed people, serotonin receptor editing is an adaptive response to reduce stress. 'The weight of evidence is that this is an adaptive response,' Schmauss told me. Pers. comm., 29 April 2011.

92 Nowacki, M., et al., 'RNA-mediated Epigenetic Programming of a Genome-Rearrangement Pathway', 2008, *Nature*, vol. 451, p. 153.

93 I modified the following excerpt, 'The ability to transmit a memory of adaptation to experience would be not only subject to natural selection, but also potentially favoured by it. Recent findings suggest that this might occur and that all of the mechanisms required are either in place or possible.' From Mattick, J.S., 'Has Evolution Learnt How to Learn?', 2009, *EMBO Reports*, vol. 10, p. 665, <nature.com/embor/journal/v10/n7/full/embor2009135.html>, viewed 27 May 2011.

94 'Office Hours With Randy Jirtle', 22 January 2010, <youtube.com/watch?v=GFK5xPhkhHM>, viewed 27 May 2011.

95 Ptashne, M., Hobert, O. and Davidson, E., 'Questions on the Scientific Basis of the Epigenome Project', 2010, *Nature*, vol. 464, p. 487.

96 Epigenetics did have one high profile setback. A study of identical twins, one of whom had multiple sclerosis, failed to show any difference in their epigenetic marks after their DNA was sequenced. The study was published in *Nature* in April 2010 by Sergio Baranzini and colleagues at the University of California, San Francisco.

97 Waterland, R.A. and Jirtle, R.L., 'Transposable Elements: Targets for Early Nutritional Effects on Epigenetic Gene Regulation', 2003, *Molecular and Cellular Biology*, vol. 23, p. 5293. And also a word of warning specifically about the taking of choline supplements. A study reported in April 2011 in *Nature* showed that gut bacteria can metabolise it to a compound that is toxic to the heart. Choline is an essential nutrient but, 'We have no idea how much is too much,' Hazen says. Many multivitamins and other dietary supplements now contain choline, but eating large amounts of the substance might not be healthy. 'Our data would suggest that's a bad thing,' he says. 'It suggests that's like eating a tub of cholesterol.' Excerpted from <sciencenews.org/view/generic/id/72372/title/Gut_microbes_may_foster_heart_disease>, viewed 27 May 2011.

Chapter 4—Your Genetic Future

1 The '23 and Me' blog is titled 'The Spittoon'.

2 The cofounder of '23 and Me', Anne Wojcicki, is married to Sergey Brin, the cofounder of Google.

3 I interviewed Joe DeRisi at a malaria conference in Lorne, Victoria, 7 February 2008.

4 Baltimore, D., 'Mapping the Genome: The Vision, the Science, the Implementation', 1992, *Los Alamos Science*, 20, p. 78:

> ... the Genome Project is something quite different because it will allow us to examine human variability, for example, variations in mathematical ability or in what we call intelligence. Those variations are caused by the interaction of many genes. And certainly the best way that biologists have to unravel which genes are involved in complex traits is to find a set of markers that are linked to the disease and then find the genes associated with those markers ... I'm very interested in studying the genome at a level where we can get at multigenic traits and at subtle aspects of human genetics ...
>
> But the only way to study the genetics of the higher perceptual and integrative human functions is by studying human beings ... So I believe that the Human

Genome Project will open up an entirely new level of human biology. To my mind that is the only reasonable rationale for the whole program.

5 Imagine a full stop as a sphere with a 0.5 millimetre diameter and a nucleus (where DNA resides) as a sphere of 5 micrometres diameter—that's a hundredfold difference in diameter, which equates to a millionfold difference in volume.

6 When a trait like height differs across a population, 'heritability' is a measure of how much of that difference is due to genes. For a tutorial on how trait heritability is defined, see Wray, N. and Visscher, P., 'Estimating Trait Heritability', 2008, *Nature Education*, 1(1), <nature.com/scitable/topicpage/estimating-trait-heritability-46889>, viewed 18 June 2011.

7 Type 2 diabetes is estimated to be between 35% and 58%. Stumvoll, M., et al., 'Type 2 Diabetes: Principles of Pathogenesis and Therapy', 2005, *Lancet*, vol. 365, p. 1333. Schizophrenia is estimated to be between 73% and 90% heritable. Sullivan, P.F., et al., 'Schizophrenia as a Complex Trait: Evidence From a Meta-analysis of Twin Studies', 2003, *Archives of General Psychiatry*, vol. 60, p. 1187. The heritability of autism is estimated to be greater than 90%; see Gupta, A.R. and State, M.W., 'Recent Advances in the Genetics of Autism', 2007, *Biological Psychiatry*, vol. 61, p. 429.

8 *Gattaca*, written and directed by Andrew Niccol, Sony Studios, 1997.

9 Madrigal, A., 'Are Internet Genetic Testing Services *Really* Illegal?', 2008, <wired.com/wiredscience/2008/06/are-genetic-tes/>, viewed 18 June 2011. Germany also passed legislation to ban direct-to-consumer genetic testing; see <scienceblogs.com/geneticfuture/2009/04/germany_to_ban_dtc_genetic_tes.php>, viewed 16 June 2011.

10 For updates on the debate over direct-to-consumer genetic testing, see Daniel MacArthur's blog at 'Wired Science', <wired.com/wiredscience/category/genetic-future>, viewed 6 July 2011.

11 'We now have discrimination down to a science': dialogue from *Gattaca*.

12 That's not to say that for some genes the results might not be useful. The jury is out for things like Alzheimer's disease and the ApoE4 gene. University of Melbourne geneticist Robert Williamson told me he could imagine that older people would act on this information in a sensible way. If you knew you were at ten times the risk (a consequence of carrying two copies of the ApoE4 gene variant) you might take extra care to make sure your affairs are in order. A more dramatic example is illustrated by the Chief Scientific Officer of DeCODE Genetics, Jeffrey Gulcher. From scanning his own genome, he learnt that he had twice the risk for prostate cancer. That result prompted his primary care physician to order a prostate specific antigen (PSA) test, which detects the presence of prostate cancer cells. Gulcher's PSA level was at the high end of normal and he wound up seeing a urologist, who diagnosed him with high-grade cancer on both sides of his prostate. The cancer was then surgically removed. See <genomeweb.com/dxpgx/science-behind-dtc-genetic-testing>, viewed 18 June 2011.

13 McCarthy, M. and Zeggini, E., 'Genome-wide Association Studies in Type 2 Diabetes', 2009, *Current Diabetes Reports*, vol. 9, p. 164.

14 Pers. comm., 21 April 2011.

15 A series of controversial articles that were published in the *New England Journal of Medicine* on 15 April 2009 are discussed on Daniel Macarthur's blog; see <scienceblogs.com/geneticfuture/2009/04/personal_genomics_is_not_dead.

php>, viewed 18 June 2011; and see also Wade, N., 'Genes Show Limited Value in Predicting Diseases', *The New York Times*, 15 April 2009, <nytimes. com/2009/04/16/health/research/16gene.html>, viewed 29 June 2011.

16 Wade, N., 'A Dissenting Voice as the Genome is Sifted to Fight Disease', *The New York Times*, 15 September 2008, <http://www.nytimes.com/2008/09/16/science/16prof.html>, viewed 29 June 2011.

17 Alleyne, R. and Devlin, K., 'Genetic "Magic Bullet" Cures Have Proven a "False Dawn"', 2009, <telegraph.co.uk/scienceandtechnology/science/sciencenews/5190914/Genetic-magic-bullet-cures-have-proven-a-false-dawn.html>, viewed 18 June 2011.

18 Interview, Eric Lander, 11 June 2009. Lander is director of the Broad Institute at Massachusetts Institute of Technology.

19 Finkel, E., <biosino.org/bioinformatics/010904-6.htm>, viewed 28 May 2011.

20 Ibid.

21 Faulkner, W., *The Sound and the Fury*, 1929, Jonathon Cape and Harrison Smith, quoted in the preface to Terwilliger, J. and Weiss, K., 'How Many Diseases Does It Take to Map a Gene With SNPs?', 2000, *Nature Genetics*, vol. 26, p. 151.

22 Six billion because each person has two copies of a genome.

23 The National Human Genome Research Institute compared Moore's law—the benchmark for rapid technological innovation—to the rate at which the cost of DNA sequencing had plummeted; see <genome.gov/27541954>, viewed 16 June 2011.

24 Mark McCarthy told me his lab was doing low pass-sequencing for about US$2,500 (four to six times coverage), which is adequate for disease studies with large numbers of people; and high accuracy 'deep' sequencing (covering the genome 30 to 40 times over) for US$5,000. David Goldstein told me deep sequencing was costing him around US$6,000 to $7,000. Interviews with both scientists conducted on 21 April 2011.

25 Like a snap-together toy, haemoglobin is comprised of alpha and beta modules, two of each. It was the beta module that carried the defect, preventing the four modules from forming their normal soluble shape.

26 The nucleotide A had been changed to a T. That changed the amino acid code from glutamate to valine. See Genetics Home Reference from the NIH, <ghr.nlm. nih.gov/condition=sicklecelldisease>, viewed 16 June 2011. The beta globin gene is 1,600 letters long; see <ornl.gov/sci/techresources/Human_Genome/posters/chromosome/hbb.shtml gene>, viewed 16 June 2011.

27 Ramsak, B., 'Event Report: Men's 100m Final', 2009 <berlin.iaaf.org/news/kind=108/newsid=52999.html>, viewed 18 June 2011.

28 The use of restriction enzymes to determine sites of DNA variation and correlate these variations with disease is attributed to two geneticists, David Botstein and Y.W. Kan; see de Souza, N., 'RFLP Realization', 2007, <nature.com/milestones/miledna/full/miledna08.html>, viewed 18 June 2011.

29 Bacteria used these so-called restriction enzymes as weapons to chop up the DNA of invading parasites; they protect their own DNA by coating the same sites with protective methyl groups.

30 Restriction enzymes provided the first way to 'read' differences in genomic DNA, but a few years later researchers graduated to using DNA 'stutters'—places where the

DNA code starting repeating itself (itself itself itself itself itself itself). Comparing these stutters in any two people, researchers would find lots of differences across the genome—far more than they would if comparing the cut sites for restriction enzyme. So it was easier to find a stutter marker that was linked to a disease gene. The stutters were spaced across the landscape of the genome like highway markers—about ten times per chromosome, or once every ten million letters. So the gene miners had it made. It was just a matter of linking a particular version of the highway marker to the disease and, presto, the coordinates of the disease gene were at hand. See Casci, T., 'Milestone 12 (1985), "DNA Fingerprinting, a Repeat Success"', 2007, <nature.com/milestones/miledna/full/miledna12.html>, viewed 16 June 2011.

31 Gusella, J.F., et al., 'A Polymorphic DNA Marker Genetically Linked to Huntington's disease', 1983, *Nature*, vol. 306, p. 234.

32 Walker, F., 'Huntington's Disease', 2007, *The Lancet*, vol. 369, p. 218.

33 Williamson, R., 'Towards a Total Human Gene Map', 1985, *The Woodhull Lecture*, Royal Institute Proceedings, p. 45.

34 The figure for 2,565 Mendelian genes comes from Ada Hamosh, the curator of the NIH's database called Online Mendelian Inheritance in Man, or OMIM. I sent her an email asking her how many Mendelian genes there are. Her answer was, 'There are 22,500 genes, we catalog over 13,000 in OMIM at this time. If you mean how many genes have mutations causing how many Mendelian disorders, the answer as of April 15, 2011 is 2565 genes with mutations causing 4321 disorders.' Ada Hamosh, MD, MPH, Clinical Director, Institute of Genetic Medicine, Scientific Director, OMIM, pers. comm., 3 May 2011.

35 Families with hypercholesterolemia were commonly found to carry a mutation in their LDL receptor gene, highlighting the role of this receptor in lowering blood cholesterol levels. Statins are drugs that lower cholesterol levels, largely by raising the levels of the LDL receptor in the liver. The effect is indirect. Statins inhibit HMG CoA reductase, which inhibits liver cholesterol synthesis. That raises the levels of LDL receptors, presumably to draw more cholesterol from the bloodstream to the liver. Rader, D.J., et al., 'Monogenic Hypercholesterolemia: New Insights in Pathogenesis and Treatment', *Journal of Clinical Investigation*, 2003, vol. 111, p. 1795.

36 Couzin-Frankel, J., 'The Promise of a Cure: 20 Years and Counting', 2009, *Science*, vol. 324, p. 1504.

37 Even though only a handful of peoples' genomes were sequenced, each genome carries two copies of the code, so it is immediately obvious when you sequence that some positions of the code are identical in both copies while other positions vary. The variable ones are the SNPs. More information on SNPs and a reference to 'one SNP every 300 letters' can be found at: <ghr.nlm.nih.gov/handbook/genomicresearch/snp>, viewed 30 June 2011.

38 The most common changes were conservative swaps, exchanging bases (letters) that were chemically similar—so, the commonly exchanged bases adenine and guanine are both purines, while the commonly exchanged bases cytosine and thymidine, are both pyrimidines.

39 '"The Future of Genetic Studies of Complex Human Diseases": Drs. Merikangas and Risch Talk About That Influential Paper 13 Years Later', <hum-molgen.org/

NewsGen/05-2009/000001.html>, viewed 30 June 2011; original paper: Risch, N. and Merikangas, K., 'The Future of Genetic Studies of Complex Human Diseases', 1996, *Science*, vol. 273, p. 1516.

40 I heard Charles Cantor make this prediction in a talk he gave at Walter and Eliza Hall Institute in Melbourne in mid-2001. I referred to it in an article I wrote a few months later; see Finkel, E., 'Genetic Pattern Hunters Are "Fishing, Not Thinking"', 2001, <biosino.org/bioinformatics/010904-6.htm>, viewed 16 June 2011.

41 It cost 20 cents to $1 per SNP per person, according to Pan Pui-Yan Kwok, quoted in Gura, T., 'Can SNPs Deliver on Susceptibility Genes?', 2001, *Science*, vol. 293, p. 593. Ten million SNP variants in a genome corresponds to a SNP every 300 bases; see <genome.gov/11511175>, viewed 16 June 2011.

42 Formally, this was the common disease–common variant hypothesis as articulated by David Reich and Eric Lander; see Reich, D.E. and Lander, E.S., et al., 'On the Allelic Spectrum of Human Disease', 2001, *Trends in Genetics*, vol. 17, p. 502.

43 Mark Daly, a geneticist at Harvard University, Boston, discovered that the shuffling across the genome was not even during his studies of unrelated families that carried Crohn's disease. See Halim, N., 'Scientists Build Case for "Haplotype" Map of Human Genome', 2001, <mit.edu/newsoffice/2001/crohns-1024.html>, viewed 16 June 2011.

44 'HapMap Provides "Catalog" of Human Genetic Variation', 2005, <web.mit.edu/newsoffice/2005/hapmap.html>, viewed 16 June 2011.

45 Interview with geneticist Greg Gibson at the University of Queensland, 8 May 2009.

46 Angier, N., 'Do Races Differ? Not Really, Genes Show', 2000, *The New York Times*, F1, <nytimes.com/2000/08/22/science/do-races-differ-not-really-genes-show.html?scp=1&sq=Angier%202000%20race&st=cse&pagewanted=all>, viewed 18 June 2011.

47 Ibid.

48 DNA Learning Center, 'Eugenics, Forced Sterilization, the Holocaust and the Gene Age', <eugenicsarchive.org/eugenics/>, viewed 18 June 2011.

49 Abraham, C., 'The New Science of Race', 2005, <theglobeandmail.com/servlet/story/RTGAM.20050618.wxrace0618/BNStory/Front/>, viewed 18 June 2011.

50 The figure one in a hundred million comes from dividing the normal significance hurdle, $p < 0.05$, by three million tests (for each SNP); see '"The Future of Genetic Studies of Complex Human Diseases": Drs. Merikangas and Risch Talk About That Influential Paper 13 Years Later', <hum-molgen.org/NewsGen/05-2009/000001.html>, viewed 30 June 2011.

51 Holden, C., 'Back to the Drawing Board for Psychiatric Genetics', 2009, *Science*, vol. 324, p. 1628.
   Another example of a highly fêted association that has crumbled is between a variant of the Calpain 10 gene and type 2 diabetes. It was the first gene association fingered for type 2 diabetes, and was identified in Mexican Americans, who tend to have a high incidence of the disease. A SNP on chromosome 2 showed what was then considered a statistically significant relationship with the disease, a probability of less than one in 20 of occurring by chance (expressed as $p < 0.05$). But as Mark McCarthy, told me, 'This association has bubbled along for the last ten years and no-one can replicate it.' I interviewed Mark McCarthy at the Lorne genome

conference in February 2009. The original paper is Horikawa, Y., et al., 'Genetic Variation in the Gene Encoding Calpain-10 is Associated With Type 2 Diabetes Mellitus', 2000, *Nature Genetics,* vol. 26, p. 163.

52 New 'gene chip' technology developed by companies like Affymetrix and Illumina had reduced the costs, so they could afford to test more SNPs in larger populations.

53 Pennisi, E., 'Breakthrough of the Year: Human Genetic Variation', 2007, *Science,* vol. 318, p. 1842.

54 The Wellcome Trust Case Control Consortium, 'Genome-wide Association Study of 14,000 Cases of Seven Common Diseases and 3,000 Shared Controls', 2007, *Nature,* vol. 447, p. 661; and commentary at Bowcock, A.M., 'Genomics: Guilt by Association', 2007, *Nature,* vol. 447, p. 645.

55 Press Release, 'First Individual Diploid Human Genome Published By Researchers at J. Craig Venter Institute: Sequence Reveals That Human to Human Variation is Substantially Greater Than Earlier Estimates', 2007, <jcvi.org/cms/press/press-releases/full-text/article/first-individual-diploid-human-genome-published-by-researchers-at-j-craig-venter-institute/> and <genome.gov/10001551#1>, viewed 18 June 2011. See also Redon, R., et al., 'Global Variation in Copy Number in the Human Genome', 2006, *Nature,* vol. 444, p. 444.

56 Choi, C., 'Starchy Diet Boosts Gene Copy Number', 2007, <the-scientist.com/news/display/53576/>, viewed 18 June 2011.

57 The NOD2 gene had been linked to Crohn's disease from studies in families in 2001; see Hugot, J.P., 'Association of NOD2 Leucine-rich Repeat Variants With Susceptibility to Crohn's Disease', 2001, *Nature,* vol. 411, p. 599; and Ogura, Y., et al., 'A Frameshift Mutation in *NOD2* Associated With Susceptibility to Crohn's Disease', 2001, *Nature,* vol. 411, p. 603. However, the fact that it emerged as a major player in the genome-wide association studies means it was also a major contributor to Crohn's disease in the general population, not just in isolated families. Sometimes a gene that causes disease in families does not seem to be associated with the disease in the general population. BRCA1, for instance, is a cause of breast cancer in families yet has not been identified in genome-wide association studies of population breast cancer. The Wellcome Trust Case Control Consortium, 'Genome-wide Association Study of 14,000 Cases of Seven Common Diseases and 3,000 Shared Controls', 2007, *Nature,* vol. 447, p. 661.

58 Hugot, J-P., et al., 'Prevalence of CARD15/NOD2 Mutations in Caucasian Healthy People', 2007, *The American Journal of Gastroenterology,* vol. 102, p. 1259.

59 Hedl, M., et al., 'Chronic Stimulation of Nod2 Mediates Tolerance to Bacterial Products', 2007, *Proceedings of the National Academy of Science,* vol. 104, p. 19440.

60 Hugot, J-P., et al., 'Crohn's Disease: The Cold Chain Hypothesis', 2003, *Lancet,* vol. 362, p. 2012.

61 Svoboda, E., 'The Worms Crawl In', 2008, <nytimes.com/2008/07/01/health/research/01prof.html?_r=2&pagewanted=1&sq&st=nyt&scp=2>, viewed 18 June 2011.

62 Hageman, G.S., 'An Integrated Hypothesis That Considers Drusen as Biomarkers of Immune-mediated Processes at the RPE-Bruch's Membrane Interface in Aging and Age-related Macular Degeneration', 2001, *Progress in Retinal and Eye Research,* vol. 20, p. 705.

63 Daiger, S.P., 'Was the Human Genome Project Worth the Effort?', 2005, *Science,* vol. 308, p. 362; and Scholl, H.P., 'An Update on the Genetics of Age-related

Macular Degeneration', 2007, *Molecular Vision*, vol. 13, p. 196. The company Optherion also has a good discussion on its website; see <optherion.com/our-technologies/the-complement-system-and-genetics-of-disease>, viewed 18 June 2011.

64 Type 2 diabetes caused by a single dominant gene is a rare condition known as Mature Onset Diabetes of the Young, or MODY. It runs in families and studies of different families identified six genes that can individually cause the condition. Two of them, HFN1B (TCF2) and KCNJ1, also play a role in common type 2 diabetes. See Fajans, S., et al., 'Molecular Mechanisms and Clinical Pathophysiology of Maturity-Onset Diabetes of the Young', 2001, *New England Journal of Medicine,* vol. 345, p. 971.

65 As explained to me by Mark McCarthy of Oxford University, the lead researcher behind the type 2 diabetes genome-wide association scans. Pers. comm., 12 and 21 April 2011.

66 'Major International Collaboration Offers New Clues to Genetics of Type 2 Diabetes', 2008, <wellcome.ac.uk/News/Media-office/Press-releases/2008/WTD 039333.htm>, viewed 18 June 2011.

67 Neuregulin was first identified by DeCODE Genetics in Icelandic people with schizophrenia and later in Scottish sufferers. DISC was discovered in a large Scottish family, and Disbindin in Irish, German, British and Bulgarian families. For a review, see Owen, M.J., et al., 'Schizophrenia: Genes at Last?', 2005, *Trends in Genetics,* vol. 21, p. 518; and Hamilton, S.P., 'Schizophrenia Candidate Genes: Are We Really Coming Up Blank?', 2008, *American Journal of Psychiatry,* vol. 165, p. 420.

68 Wade, N., 'Genes May Play a Role in Schizophrenia', *The New York Times,* 13 December 2002; <nytimes.com/2002/12/13/national/13GENE.html>, viewed 18 June 2011.

69 The prevalence of schizophrenia varies across different populations from 0.3% to 0.7%, according to a recent study. van Os, J., et al., 'Schizophrenia', 2009, *Lancet,* vol. 374, p. 635.

70 Walsh, T., et al., 'Rare Structural Variants Disrupt Multiple Genes in Neurodevelopmental Pathways in Schizophrenia', 2008, *Science,* vol. 320, p. 539; and Holden, C., 'Rare Mutations Hint at Multiple Schizophrenias', 2008, <sciencenow.sciencemag.org/cgi/content/full/2008/327/1>, viewed 18 June 2011.

71 David Goldstein, pers. comm., 4 May 2011, regarding the ability to track down the gene that is associated with a disease:

> The track record here has been poor, most signals are not resolved to causal variants, and this means normally one does not even know what gene is involved; this compromises the value substantially since it is hard to learn much about the biology or to identify a new therapeutic target if you don't have the gene.

72 Phillips, N., 'Genetic Link Between MS and Vitamin D', 2009, <abc.net.au/science/articles/2009/06/15/2596859.htm>, viewed 18 June 2011.

73 HLA genes play a key role in the immune system and are linked to many autoimmune diseases.

74 Simon Foote, pers. comm., May 2009.

75 Couzin, J. and Kaiser, J., 'Closing the Net on Common Disease', 2007, *Science,* vol. 316, p. 82.

76 McCarthy, M.I. and Zeggini, E., 'Genome-wide Association Studies in Type 2 Diabetes', 2009, *Current Diabetes Reports*, vol. 9, p. 164.

77 Maher, B., 'Personal Genomes: The Case of the Missing Heritability', 2008, <nature.com/news/2008/081105/full/456018a.html>, viewed 18 June 2011.

78 Weedon, M.N., et al., 'Genome-wide Association Analysis Identifies 20 Loci That Influence Adult Height', 2008, *Nature Genetics*, vol. 40, p. 575.

79 Kathiresan, S., et al., 'Polymorphisms Associated With Cholesterol and Risk of Cardiovascular Events', *New England Journal of Medicine*, 2008, vol. 358, p. 1240; and Myocardial Infarction Genetics Consortium, 'Genome-wide Association of Early-onset Myocardial Infarction With Common Single Nucleotide Polymorphisms, Common Copy Number Variants, and Rare Copy Number Variants', 2009, *Nature Genetics*, vol. 41, p. 334.

80 The International Schizophrenia Consortium, 'Common Polygenic Variation Contributes to Risk of Schizophrenia and Bipolar Disorder', 2009, *Nature*, vol. 460, p. 748.

81 Stefansson, H., et al., 'Common Variants Conferring Risk of Schizophrenia', 2009, *Nature*, vol. 460, p. 744.

82 'Guest Post: Neil Walker on the Curious Case of the Schizophrenia GWAS', 2009, <scienceblogs.com/geneticfuture/2009/07/guest_post_neil_walker_on_the.php#moren>, viewed 18 June 2011.

83 Wade, N., 'Hoopla, and Disappointment, in Schizophrenia Research', July 2009, <tierneylab.blogs.nytimes.com/2009/07/01/hoopla-and-disappointment-in-schizophrenia-research/>, viewed 3 July 2011.

84 Wade, N., 'A Dissenting Voice as the Genome Is Sifted to Fight Disease', *The New York Times*, 2008, p. F3, <nytimes.com/2008/09/16/science/16prof.html>, viewed 10 July 2011.

85 There were exceptions. Genome-wide scans found a large fraction of the genes that contribute to type 1 diabetes, macular degeneration and blood lipid levels. See Lander, E.S., 'Initial Impact of the Sequencing of the Human Genome', 2011, *Nature*, vol. 470, p. 187.

86 From Mauricio, R., 'Mapping Quantitative Trait Loci in Plants: Uses and Caveats for Evolutionary Biology', 2001, *Nature Reviews Genetics*, vol. 2, p. 370, Table 1:

The early history of evolutionary genetics focused on understanding complex traits, particularly those relating to humans, such as intelligence, temper and 'artistic faculty' ... With the rediscovery of Mendel's theory of heredity in 1900, a conflict arose between this 'biometrical' school of quantitative genetics and the discrete genetics of the Mendelian school. By 1910, it had been shown that continuous phenotypic variation could result from the action of the environment on the segregation of many Mendelian loci. By 1918, Ronald Fisher convincingly reconciled the discrete inheritance of Mendelism with the biometrical approach. Modern evolutionary quantitative genetics is largely based on the same statistical foundations that were laid by Pearson and Fisher. For most of the twentieth century, quantitative genetics had a crucial role in both agriculture and evolutionary biology, but never seemed fully embraced by modern molecular genetics. There was (and perhaps still is) a widespread perception that quantitative genetics essentially ignored genetics, blanketing actual genes in what has been called a 'statistical fog'.

87 Modern geneticists began referring to the problem of dark matter as the elegantly phrased 'architecture of complex traits'. Here 'complex' is used to distinguish them from Mendelian traits, which are caused by a single gene. Shao, H., et al., 'Genetic Architecture of Complex Traits: Large Phenotypic Effects and Pervasive Epistasis', 2008, *Proceedings of the National Academy of Sciences USA*, vol. 105, p. 19910. For a scholarly analysis of 'dark matter' by the gurus of the field, see Manolio, T.A., et al., 'Finding the Missing Heritability of Complex Diseases', 2009, *Nature*, vol. 461, p. 747.

88 Finkel, E., 'Genes for Height Hiding in Plain Sight', 2010, <news.sciencemag. org/sciencenow/2010/06/genes-for-height-hiding-in-plain.html>, viewed 18 June 2011; Lander, E.S., 'Initial Impact of the Sequencing of the Human Genome', 2011, *Nature*, vol. 470, p. 187.

89 Allen, H.L., et al., 'Hundreds of Variants Clustered in Genomic Loci and Biological Pathways Affect Human Height', 2010, *Nature*, vol. 467, p. 832.

90 Scaling up does not always not always make a dent in the dark matter. Another gargantuan study was carried out for type 2 diabetes. They found another 12 SNPs, bringing the total to 38. But this was a case of diminishing returns. The extra genes had such tiny effects that overall the 38 genes still explain only 5–10% of the genetic heritability of the disease; see Voight, B.F., et al., 'Twelve Type 2 Diabetes Susceptibility Loci Identified Through Large-scale Association Analysis', 2010, *Nature Genetics*, vol. 42, p. 579. Scaling up the operation for Crohn's disease also produced very small returns. Until 2008, they had found 32 loci (gene markers) that explained 20% of the heritable component of Crohn's disease. The scaled-up study added 39 more genes but added only an extra 3.2%. Altogether, the 71 genes explain 23.2% of the heritable component of Crohn's disease; see Franke A., et al., 'Genome-wide Meta-analysis Increases to 71 the Number of Confirmed Crohn's Disease Susceptibility Loci', 2010, *Nature Genetics*, vol. 42, p. 1118.

91 Pers. comm., 3 May 2011.

92 McClellan, J. and King, M-C., 'Genetic Heterogeneity in Human Disease', 2010, *Cell*, vol. 141, p. 210.

93 David Goldstein, pers. comm., April 2011.

94 Boettcher, P.J. and Hoffmann, I., 'Livestock Genomics in Developing Countries', *Science*, 2009, vol. 324, p. 1515; and Hayes, B.J., 'Genomic Selection in Dairy Cattle: Progress and Challenges', 2009, *Journal of Dairy Science*, vol. 92, p. 433.

95 Pers. comm., August 2009.

96 Shao, H., et al., 'Genetic Architecture of Complex Traits: Large Phenotypic Effects and Pervasive Epistasis', 2008, *Proceedings of the National Academy of Science USA*, vol. 105, p. 19910.

97 As explained to me by Georgia Chenevix-Trench, a geneticist at the Queensland Institute of Medical Research. There is a global consortium of researchers that goes by the name of CIMBA (Consortium of Investigators of Modifiers of BRCA1/2) that is searching for genes that modify the risk of carrying a BRCA1/2 gene. Similar findings have been made for cystic fibrosis. Though the disease is caused by inheriting a mutation at the CFTR gene, the severity of the disease depends on what other genes are co-inherited; see Guggino, W.B. and Stanton, B.A., 'New Insights Into Cystic Fibrosis: Molecular Switches That Regulate CFTR', 2006, *Nature Reviews Molecular Cell Biology*, vol. 7, p. 426. For a nice tutorial on Mendelian diseases, see Heidi, C.,'Rare Genetic Disorders: Learning

About Genetic Disease Through Gene Mapping, SNPs, and Microarray Data',
2008, *Nature Education*, vol. 1(1); e-document available at <nature.com/scitable/
topicpage/rare-genetic-disorders-learning-about-genetic-disease-979>, viewed 18
June 2011.

98 If there are a thousand interacting genes that can cause a disease like schizophrenia,
and there are two mutations in each gene that change how it interacts, we'd need
2^{1000} people to figure out how to predict schizophrenia. Mark McCarthy, pers.
comm., April 2001.

99 See <genomics.xprize.org/archon-x-prize-for-genomics/prize-overview>, viewed 18
July 2011.

The X-PRIZE requires that the company sequence 100 individual genomes
with an accuracy of more than 99% within ten days. Each sequence must include
at least 98% of a genome and cost $10,000 or less. The Californian company
Complete Genomics is very close. According to an email from Jennifer Turcotte,
Vice-President of Marketing at Complete Genomics, the company is able to do the
last two things, but not within ten days:

> Projects of 8 genomes or more start at $9,500/sample. This is 40x coverage. It does
> pick up fine mutations. We offer volume-based pricing whereby larger projects have
> a lower price per sample. Complete Genomics cannot accomplish sequencing of a
> complete human genome within 10 days, nor can any other vendor in the market
> today. And those who are doing sequencing in a short period of time aren't doing the
> complete human genome.

Pers. comm., 15 April 2011.

100 Walsh, F., 'Era of Personalised Medicine Awaits', 2009, <news.bbc.co.uk/2/hi/
health/7954968.stm>, viewed 18 June 2011.

101 'The genotyping process we use analyzes nearly 1,000,000 locations in a person's
genome', <spittoon.23andme.com/>; and for the latest price of a '23 and Me' test,
see <23andme.com/>, viewed 6 July 2011.

102 Dr Church even argues that genome sequencing 'will in effect be available free'
because companies will give away sequencing to sell other services, such as genetic
interpretation—much as mobile operators 'give away' handsets to get customers to
sign up for lucrative service plans. And when this happens, he reckons, 'It will be
just like the internet: once all this information is floating around, a lot of creative
people with PCs will nose around and develop applications.' 'Getting Personal',
16 April 2009, *The Economist*, <economist.com/node/13437974>, viewed 6 July
2011; and Dolan, B., 'Apple Sheds Light on Illuminas Planned Genome App', 2009,
<mobihealthnews.com/4358/apple-sheds-light-on-illuminas-planned-genome-
app/>, viewed 18 June 2011.

103 Interview, Eric Lander, May, 2009.

Chapter 5—Surviving AIDS

1 Two groups discovered the virus in 1983. Robert Gallo's group at the US National
Cancer Institute named it human T-cell leukaemia virus III (HTLV III), based
on the similarity of its shape to other HTLVs his group had isolated. Luc
Montagnier's group at the Pasteur Institute in France named it lymphadenopathy-
associated virus (LAV) based on the fact his group isolated it from a patient with

lymphadenopathy, or swollen lymph nodes. The name was later changed to HIV. Gallo, R.C., et al., 'Isolation of Human T-cell Leukemia Virus in Acquired Immune Deficiency Syndrome (AIDS)', 1983, *Science*, vol. 220, p. 865; Barre-Sinoussi, F., et al., 'Isolation of a T-lymphotropic Retrovirus From a Patient at Risk for Acquired Immune Deficiency Syndrome (AIDS)', 1983, *Science*, vol. 220, p. 868.

Puzzlingly, Montagnier won a Nobel Prize for his work, but Gallo was left out; see Cohen, J. and Enserink, M., 'HIV, HPV Researchers Honored, But One Scientist Is Left Out', 2008, *Science*, vol. 322, p. 174.

2 In Australia in 2009, there were 1,050 new cases of HIV. Sixty-five per cent were men who have sex with men, 28.7% were infected through heterosexual contact, 2.3% were injecting drug users, and 3% were men who both have sex with men and are injecting drug users. Thirteen per cent were women. Overall, there are 29,395 people living with HIV in Australia; see <avert.org/aids-hiv-australia.htm>, viewed 30 May 2011.

3 It's not completely normal; they have a 50% chance of making it to their seventieth birthday. Estimate based on interviews with Sharon Lewin, Director of the Infectious Diseases Unit at the Alfred Hospital in Melbourne, Australia, during July 2010; and Lohse, N., et al., 'Survival of Persons With and Without HIV Infection in Denmark, 1995–2005', 2007, *Annals of Internal Medicine*, vol. 146, p. 87.

4 Worldwide, 33.3 million people are infected, 22.5 million of them in Africa; see <unaids.org/globalreport/Global_report.htm>, viewed 30 May 2011.

5 Ibid.

6 The prevalence of HIV in adults (people aged 15 to 49 years) in Botswana is now 24.8%, slightly down from 2001, when it held the world record for HIV prevalence at 26.3%. Swaziland now holds that distinction with a prevalence of 25.9%, up from 23.6% in 2001; see p. 181 of <unaids.org/globalreport/Global_report.htm>, viewed 30 May 2011.

7 <unaids.org/globalreport/Global_report.htm>, viewed 30 May 2011.

8 An estimated 80% of the infected people in sub-Saharan Africa do not know their HIV status. In Botswana, the figure is 60%. In the US, by contrast, about 25% of infected people are unaware they are infected. According to one model, each untreated infected person infects seven others before they die; statistic reported by Cohen, J., 'Treat Everyone Now? A "Radical" Model to Stop HIV's Spread', 2008, *Science*, vol. 322, p. 1453. The estimate of 60% for Botswana comes from 'Botswana AIDS Impact Survey 3, Preliminary Results', 2009: 'Percentage of women and men aged 15–49 who received an HIV test in the last 12 months and who know their results is 40%'.

9 O'Brien, S.J., *Tears of the Cheetah*, 2003, St Martin's Press, New York, p. 224.

10 Newman, M.E.J. and Palmer, R.G., *Modelling Extinction*, 2003, Oxford University Press, New York, p. 1. The authors estimate that 50 million species remain out of an estimated 4 billion.

11 <who.int/gb/ebwha/pdf_files/WHA59/A59_9-en.pdf>, viewed 8 July 2011.

12 <ncbi.nlm.nih.gov/pubmedhealth/PMH0001167/>, viewed 8 July 2011.

13 <thieme-connect.de/DOI/DOI?10.1055/s-2007-996025>, viewed 8 July 2011.

14 Kerlin, B.A., et al., 'Survival Advantage Associated With Heterozygous Factor V Leiden Mutation in Patients With Severe Sepsis and in Mouse Endotoxemia', 2003, *Blood*, vol. 102, p. 3085.

15 See, for a discussion of heterozygote advantage, <en.wikipedia.org/wiki/ Heterozygote_advantage>, viewed 30 May 2011.

16 The normal virus produces 'env', a protein that binds a receptor on the cell to gain entry. The pruned virus that had taken up residence in the Lake Casitas mice released its own env protein, which clogged up the receptor, effectively stopping new viruses from infecting the cells; see O'Brien, S.J., *Tears of the Cheetah*, 2003, St Martin's Press, New York, p. 9.

17 This estimate has plummeted over the past 20 years. It is now considered that humans carry 22,500 genes. Ada Hamosh, Clinical Director, Institute of Genetic Medicine, Scientific Director, OMIM, pers. comm., 3 May 2011.

18 The CCR5 receptor was found because researchers found that molecules called cytokines (in particular Rantes, MIP-1 alpha, MIP-1 beta) could block the entry of HIV into macrophages. They guessed the reason was the cytokines were competing with HIV for a common receptor. In searching for that receptor, the researchers were led to CCR5.

19 O'Brien, S.J., *Tears of the Cheetah*, 2003, St Martin's Press, New York, p. 209.

20 Dean, M., et al., 'Genetic Restriction of HIV-1 Infection and Progression to AIDS by a Deletion Allele of the CKR5 Structural Gene', 1996, *Science*, p. 1856; Liu, R., et al., 'Homozygous Defect in HIV-1 Coreceptor Accounts for Resistance of Some Multiply-exposed Individuals to HIV-1 Infection', 1996, *Cell*, p. 367; Samson, M., et al., 'Resistance to HIV-1 Infection in Caucasian Individuals Bearing Mutant Alleles of the CCR-5 Chemokine Receptor Gene', *Nature*, 1996, p. 722.

21 O'Brien has about four hundred people in his study group who have been repeatedly exposed to HIV, usually through high-risk sex, but who have never become infected. Of these, 17 carry the double dose of the defective CCR5 receptor. Genetic tests have so far been unable to explain why the other uninfected people have been protected. Pers. comm., 8 July 2008.

22 In rare cases, people with the CCR5 Δ32 gene have been infected by strains of the virus that use alternative receptors. At late stages of the infection, HIV usually switches to using the CXCR4 receptor found on T-cells, so-called T-topic strains; and these people have been infected with the T-tropic strain; see O'Brien, S.J., *Tears of the Cheetah*, 2003, St Martin's Press, New York, chapter 12.

23 O'Brien, S.J., *Tears of the Cheetah*, 2003, St Martin's Press, New York, p. 234.

24 Ibid., p. 230.

25 Ibid., p. 238.

26 Was the CCR5 Δ32 mutation selected by the medieval plagues? There are two arguments against that view. One is that researchers discovered that DNA extracted from skeletons well over 2,900 years old already carried the mutation; see Hummel, S., et al., 'Detection of the CCR5 Δ32 HIV Resistance Gene in Bronze Age Skeletons', 2005, *Genes and Immunity*, vol. 6, p. 371; and Lidén, K., et al., 'Pushing It Back: Dating the CCR5 Δ32 bp Deletion to the Mesolithic in Sweden and Its Implications for the Meso/Neo Transition', 2006, *Documenta Praehistorica*, vol. 33, p. 29.

So the mutation was clearly not uncommon well before medieval plagues ravaged the human population. O'Brien speculates that it could have been a more ancient plague or another disease that may have been responsible for selecting the CCR5 mutation; see *Tears of the Cheetah*, 2003, St Martin's Press, New York,

p. 239. However, Pardis Sabeti at Harvard's Broad Institute believes the CCR5 Δ32 mutation spread through human populations as a neutral passenger rather than through selection. She developed a technique to identify mutations that have been under rapid selection. Essentially, genes under selection spread through populations more rapidly than can be explained by random processes. Her analysis shows that the rate of CCR5 spread is more consistent with a random process. In other words, it does not look as if the mutation conferred either an advantage or disadvantage; it arose by chance, and then simply went along for the ride. For more, see Sabeti, P.C., et al., 'The Case for Selection at CCR5-Δ32', 2005, *Plos Biology*, <plosbiology.org/article/info%3Adoi%2F10.1371%2Fjournal.pbio.0030378>, viewed 8 July 2011. And for Stephen O'Brien's arguments in favour of selection, see *Tears of the Cheetah*, 2003, St Martin's Press, New York, p. 235.

27 The hope was that Selzentry™ would block the spread of the virus in an infected person. It does the job well when HIV is relying on the CCR5 'elevator' for entry into CD4 T-cells and macrophages. The problem is that late in the course of infection the virus can mutate to use a different elevator known as CXCR4. If a patient harbours virus that can use the alternate CXCR4 elevator, then Selzentry™ is not effective at stopping the spread. For this reason, patients have to be tested to see whether they already carry CXCR4 virus before being prescribed the drug. As explained to me by Gilda Tachedjian, the head of the Molecular Interactions Group at the Burnet Institute in Melbourne; pers. comm., July 2010.

For a review of clinical experience with Selzentry™, see Westby, M. and van der Ryst, E., 'CCR5 Antagonists: Host-targeted Antiviral Agents for the Treatment of HIV Infection', 2010, *Antiviral Chemistry and Chemotherapy*, vol. 20, p. 179.

28 These chemokines go by the name of Rantes and MIP-1 alpha. The more copies of the gene for MIP-1 alpha (aka CCL3L1) the slower the progression to AIDS, because both Rantes and MIP-1 alpha block the CCR5 receptor. SDF-1 binds to a different HIV receptor on helper T-cells, the one known as CXCR4. So, high levels of SDF-1 block access of HIV to these T-cells.

29 Six HLA proteins make up the uniform. They are encoded by the genes HLA A, HLA B and HLA C genes. Each gene is present in two copies, one from Mum; one from Dad.

30 Tuberculosis is an example of disease caused by a bacterium that invades cells. The malaria parasite also invades cells.

31 Of all our genes, the three HLA genes differ most from person to person. There are hundreds of varieties of them.

32 Carrington, M., et al., '*HLA* and HIV-1: Heterozygote Advantage and B^*35-Cw^*04 Disadvantage', 1999, *Science,* vol. 283, p. 1748.

33 The Nef gene of HIV interferes with the production of HLA proteins.

34 The probe is called killer immunoglobulin-like receptor, or KIR.

35 Altfeld, M. and Goulder, P., '"Unleashed" Natural Killers Hinder HIV', 2007, *Nature Genetics*, vol. 39, p. 708.

36 The editor protein is called Apobec3G.

37 For more on RNA editing, see chapter 3.

38 The human Apobec3G protein actually stows away inside the virus capsule, furiously editing the newly minted copies of the virus by changing the letter cytidine to uridine. For more about Vif and Apobec3G, see Turelli, P. and Trono,

D., 'Editing at the Crossroad of Innate and Adaptive Immunity', 2005, *Science*, vol. 307, p. 1061.

39 The faulty tag is a ubiquitin molecule, the universally recognised tag for proteins that are about to be trashed.

40 This script is suggested for people who progress more slowly to AIDS and carry SNP misspellings in their Apobec3G editor gene to prevent it being tagged by Vif; see O'Brien, S.J. and Nelson, G.W., 'Human Genes That Limit AIDS', 2004, *Nature Genetics*, vol. 36, p. 565; and Heeney, J.L., et al., 'Origins of HIV and the Evolution of Resistance to AIDS', 2006, *Science,* vol. 313, p. 462.

41 Cyclophilins are host proteins used by HIV to cushion its genome. Berthoux, L., et al., 'Cyclophilin A is Required for TRIM5α-mediated Resistance to HIV-1 in Old World Monkey Cells', 2005, *Proceedings of the National Academy of Science USA*, <ncbi.nlm.nih.gov/pmc/articles/PMC1239943/>, viewed 30 May 2011.

42 The Asian rhesus macaque is resistant to HIV but not to SIV; see Goff, S.P., 'HIV: Replication Trimmed Back', 2004, *Nature*, vol. 427, p. 791.

43 One theory for why our Trim5 alpha gene is so clumsy against HIV is that its sights are trained against a different virus that once threatened mankind—Pan troglodytes endogenous retrovirus or PtERV. PtERV breached the defences of our chimp and gorilla cousins to take up residence in their genomes but it never made it into the human genome. Human Trim5 alpha may be the reason why; it does a great job of trimming the cushions off PtERV. However, it seems these cushion trimmers can only do one thing at a time. Human Trim5 alpha is puny against HIV. Likewise, the rhesus monkey's trimmer does a great job against HIV but is not much good against PtERV. As explained to me by Michael Emerman at the Fred Hutchison Cancer Research Centre in Seattle, pers. comm., 5 August 2010.

    For a scientific account of this story, see Kaiser, S.M., et al., 'Restriction of an Extinct Retrovirus by the Human TRIM5a Antiviral Protein', 2007, *Science*, vol. 316, p. 1756; and for an enthralling popular description, see Michael Specter's article in *The New Yorker*, <newyorker.com/reporting/2007/12/03/071203fa_fact_spectre>, viewed 30 May 2011.

44 Nakajima, T., et al., 'Impact of Novel TRIM5alpha Variants, Gly110Arg and G176del, on the Anti-HIV-1 Activity and the Susceptibility to HIV-1 Infection', 2009, *AIDS*, vol. 23, p. 2091.

45 O'Brien, S.J. and Nelson, G.W., 'Human Genes That Limit AIDS', 2004, *Nature Genetics*, vol. 36, p. 565.

46 Rhodes, D., et al., 'Characterization of Three *nef*-Defective Human Immuno-deficiency Virus Type 1 Strains Associated With Long-Term Nonprogression', 2000, *Journal of Virology*, vol. 74, p. 10581.

47 Pers. comm., December 2008.

48 <content.thehurtlocker.com/20100311/index.html>, viewed 30 May 2011.

49 HIV makes a specialty of infiltrating CD4 T-cells, also known as 'helper T-cells'. They are called helpers because after they detect infected cells, they rouse other contingents of the immune system. For instance, they stimulate antibody-making B-cells and killer T-cells to proliferate.

50 <avert.org/aids-history-86.htm>, viewed 8 July 2011.

51 There is one bright spot on the horizon. After searching the blood of 1,800 people, Dennis Burton, scientific director of the IAVI Neutralizing Antibody Center at The

Scripps Research Institute in La Jolla, and his colleagues from the International Aids Vaccine Initiative (IAVI) consortium reported in September 2009 finding some extraordinary antibodies in a South African patient. These 'broadly neutralising' antibodies were able to thwart the stealth tactics of HIV. Even as HIV changed its spots, they maintained their ability to recognise the virus.

How did these antibodies overwhelm the virus? They recognise a transient three-pronged docking structure created by the viral GP120 and GP41 glycoproteins. This 'trimeric' structure is highly unstable, so most antibodies don't lock onto it, rather they lock onto a trashed form of the trimer. That sends the antibodies off on a wild goose chase. The elite antibodies that recognised the functional trimer structure evolved in the patient after years of battling HIV. 'Evolve' is the right word, because the presence of the virus unwittingly selects ever more effective antibodies. Burton describes this evolutionary race between the host's antibodies and the virus as 'a clash of Titans'.

Unfortunately for the South African patient, by the time he had evolved these elite antibodies, it was too late. By then, the virus had already depleted his immune army and made it too difficult for a few elite antibodies to do the job. However, if elite-trained antibodies like this were around from the beginning that might be a different story. The goal of the IAVI consortium is to teach the immune system how to make elite antibodies. A vaccine that presented the functional trimer to the immune system might do the job. For more information, see Burton, D.R., et al., 'Antibody vs. HIV in a Clash of Evolutionary Titans', 2005, *Proceedings of the National Academy of Science USA*, vol. 102, p. 14943, <pnas.org/content/102/42/14943.full#ref-3>, viewed 30 May 2011; Cohen, J., 'Potent HIV Antibodies Spark Vaccine Hopes', 2009, *Science*, vol. 325, p. 1195; <sciencedaily.com/releases/2009/09/090903163730.htm>, viewed 30 May 2011.

52 A recessed part of the GP120 hook also serves as the docking site with the cell's CCR5 co-receptor. This part of the virus can't afford to mutate, but it is protected from the reach of antibodies because it is only exposed once the virus has latched onto the cell's CD4 receptor. Chen, B., et al., 'Structure of an Unliganded Simian Immunodeficiency Virus gp120 core', 2005, *Nature*, vol. 433, p. 834.

53 The three HIV genes engrafted into adenovirus 5 were gag, nef and pol.

54 The STEP/Merck vaccine trial showed that patients who had previously been exposed to adenovirus 5 (as judged by high levels of antibodies to the virus) were at higher risk of being infected by HIV. One explanation was that the pre-exposed group reacted to the vaccine by mobilising T-cells that migrate to mucosal surfaces lining the vagina and anus. Having more T-cells is good—it's what you want from a vaccine. But maybe having more helper T-cells lining the anus and vagina is bad, because these are the very cells that HIV is good at infecting. This theory was backed by a paper published by Steven Patterson and colleagues at Imperial College in London based on isolated blood samples from people who were *not* in the STEP/Merck trial. It showed that when the blood of people with pre-exposure to adenovirus 5 was re-exposed to the virus it produced T-cells that were more likely to stick to mucosal surfaces. But a paper from Dan Barouch at Harvard and Michael Betts from the University of Pennsylvania showed that there was no evidence for a similar effect in the patients in the STEP trial. Comparing people who had high or low prior exposure to adenovirus 5 showed no difference in the

characteristics of their T-cells after vaccination with the Merck vaccine. To read this discussion in full, see <the-scientist.com/blog/display/56150/>, <nature.com/news/2009/090720/full/news.2009.707.html> and <hvtn.org/media/pr/step1207.html>, viewed 30 May 2011.

Another paper based on monkey research also suggests that a vaccine that mobilises more helper T-cells (HIV's target cell) rather than killer T-cells can help the process of HIV infection. See Staprans, S.I., et al., 'Enhanced SIV Replication and Accelerated Progression to AIDS in Macaques Primed to Mount a CD4 T Cell Response to the SIV Envelope Protein', 2004, *Proceedings of the National Academy of Science USA*, vol. 101, p. 13026.

55 The Thai trial immunised people with two different types of vaccine. AIDSVAX B/E carried the GP120 protein and was designed to trigger antibodies. ALVAC-HIV [vCP1521] is a canary pox virus engineered to produce fragments of the HIV proteins: Gag, Pro and Env (GP120 and GP41); see Rerks-Ngarm, S., et al., 'Vaccination With ALVAC and AIDSVAX to Prevent HIV-1 Infection in Thailand', 2009, *Lancet*, vol. 361, p. 2209, <content.nejm.org/cgi/content/full/NEJMoa0908492>, viewed 30 May 2011.

For discussion of the trial results, see Cohen, J., 'Surprising AIDS Vaccine Success Praised and Pondered', 2009, *Science*, vol. 326. p. 26; and Cohen, J., 'Beyond Thailand: Making Sense of a Qualified AIDS Vaccine "Success"', 2009, *Science*, vol. 326, p. 652.

56 Pers. comm., 23 May 2008.

57 Pers. comm., 23 May 2008.

58 Globally 36% of adults in need of AIDS treatment receive it; for children the figure is 28%; see <unaids.org/en/resources/presscentre/pressreleaseandstatementarchive/2011/june/20110603praids30/>, viewed 8 July 2011.

59 Hecht, R., et al., 'Critical Choices in Financing the Response to the Global HIV/AIDS Pandemic', 2009, *Health Affairs*, vol. 28, p. 1591.

60 As estimated by Sharon Lewin, Director of the Infectious Diseases Unit at the Alfred Hospital in Melbourne. In 2009, there were 2.6 million new HIV infections; see <usaid.gov/our_work/global_health/aids/News/aidsfaq.html>, viewed 30 May 2011.

61 'The field of vaccinology began in ignorance,' reads the opening line of this review: Pantaleo, G. and Koup, R.A., 'Correlates of Immune Protection in HIV-1 Infection: What We Know, What We Don't Know, What We Should Know', 2004, *Nature Medicine*, vol. 10, p. 806.

62 Bruce Walker is Director of the Ragon Institute of Massachusetts General Hospital, Massachusetts Institute of Technology and Harvard; and an Adjunct Professor at the Nelson Mandela School of Medicine in Durban. His research institutes in Boston and Durban carry out HIV research in labs situated a stone's throw from where patients are being treated. The Durban institute was also the first in Africa devoted to biomedical research. Predominantly staffed by African scientists and doctors, it built a new expertise among South Africans as frontline players in the battle against HIV.

63 Walker made these remarks in '25 Years of AIDS' by Michael Hirso, the Spring 2006 edition of the MGH magazine *Proto*, <protomag.com/assets/25-years-of-aids?format=print>, viewed 30 May 2011.

64 It started with a single case, known as the Berlin patient; see Lisziewicz, J., et al., 'Control of HIV Despite the Discontinuation of Antiretroviral Therapy', 1999, *New England Journal of Medicine*, vol. 340, p. 1683.

65 Gulick, R.M., 'Structured Treatment Interruption in Patients Infected With HIV: A New Approach to Therapy', 2002, *Drugs*, vol. 62, p. 245.

66 Pantaleo, G. and Koup, R.A., 'Correlates of Immune Protection in HIV-1 Infection: What We Know, What We Don't Know, What We Should Know', 2004, *Nature Medicine*, vol. 10, p. 806.

67 Walker, B.D., et al., 'HIV-specific Cytotoxic T Lymphocytes in Seropositive Individuals', 1987, *Nature*, vol. 328, p. 345.

68 Fraser, C., et al., 'Variation in HIV-1 Set-point Viral Load: Epidemiological Analysis and an Evolutionary Hypothesis', 2007, *Proceedings of the National Academy of Science USA*, vol. 104, p. 17441.

69 The figure of one in 200 people being elite controllers is based on data from an American military cohort; see Okulicz, Jason F., et al., 'Clinical Outcomes of Elite Controllers, Viremic Controllers, and Long-Term Nonprogressors in the US Department of Defense HIV Natural History Study', 2009, *The Journal of Infectious Diseases*, vol. 200, p. 1714; a summary is available at <hivandhepatitis. com/recent/2010/012209_c.html>, viewed 8 July 2011.

    While elite controllers are at a vastly reduced risk of developing AIDS, some do succumb. Steve Deeks, an HIV clinician and researcher at the University of California, San Francisco, reported in an article in *Science* magazine that four of 58 elite controllers he studied progressed to AIDS. 'If I were an elite controller, I'd seriously think about going on treatment,' he said. Cohen, J., 'HIV/AIDS Researchers Reach for High-Hanging Fruit', 2009, *Science*, vol. 323, p. 996.

70 Pers. comm., 17 July 2008.

71 Pers. comm., 12 April 2011. Walker described the difficulties he had getting the controllers study up and running:

> Like most of the better ideas I have had in my career, I was initially not able to get this funded through traditional sources. We first needed to find HIV controllers, which was no easy task. By contacting AIDS advocacy groups and HIV physicians across the US, we were able to recruit persons one by one—often a doctor with a large HIV practice would have one or two of these people, so hundreds of collaborations needed to be established, and somehow it needed to be funded. When attempts to get NIH funding failed, a private philanthropist, Mark Schwartz from New York, stepped in with a $2.5m gift to get things started. Eighteen months later, I was able to get a grant of $20m from the Gates Foundation to ramp up the effort. At that point we formed an international effort to get as many patients as we could, which was called the International HIV Controllers Study. A postdoc working with me, Florencia Pereyra, took the lead on recruiting the patients, and Paul de Bakker, a postdoctoral fellow at MGH and the Broad, was recruited to a faculty position to lead the genetic analysis team.

72 Pers. comm., April 2008.

73 In statistical parlance this means going from a p value of 0.05 to a p value of 0.00000005.

74 Pers. comm., 15 April 2010.

75 Paul de Bakker, pers. comm., September 2010.

76 Migueles, S.A., et al., 'HLA B*5701 is Highly Associated With Restriction of Virus Replication in a Subgroup of HIV-infected Long Term Nonprogressors', 2000, *Proceedings of the National Academy of Science USA*, vol. 97, p. 2709.

77 The International HIV Controllers Study, 'The Major Genetic Determinants of HIV-1 Control Affect HLA Class I Peptide Presentation', 2010, *Science*, vol. 330, p. 1551. For a more popular description of the study, see <sciencenews.org/view/generic/id/65088/title/Immune_gene_variants_help_stop_HIV> or <nature.com/news/2010/101104/full/news.2010.582.html>, viewed 30 May 2011.

78 Thomas, R., et al., 'HLA-C cell Surface Expression and Control of HIV/AIDS Correlate With a Variant Upstream of *HLA-C*', 2009, *Nature Genetics*, vol. 41, p. 1290.

79 Inside the capsule or 'capsid' are two molecules of single-stranded RNA which comprises the virus genome, the reverse transcriptase enzyme needed to recode the virus genome into DNA, an enzyme called integrase inserts the virus genome into the host genome, and proteases cut out active forms of virus proteins from a larger precursor. The protein made by the gag gene is known as p 24 or capsid protein.

80 However, people carrying HLA B57 also pay a price. They are more prone to autoimmune diseases like rheumatoid arthritis. It seems their T-cells are not only more able to recognise HIV, they are also more likely to turn on their own kind as well. A report in *Nature* in May 2010 offered a hypothesis as to why good recognition of HIV-infected cells comes at the cost of being more likely to turn on your own cells. T-cells are screened during their training. To pass, they must recognise native proteins carried in the pockets of the cells' HLA uniform. But they must adhere to the uniforms rather gingerly. If they adhere too strongly they are culled, lest they turn on their own and become autoimmune T-cells. It turns out that people with the HLA B57 uniform bind a smaller selection of the body's own proteins. That also means they end up culling a smaller selection of their T-cells, so they end up with a more diverse population, some of which do a better job of recognising gag proteins—with the downside that some of them also attack their own kind; see <nature.com/news/2010/100505/full/news.2010.219.html?s=news_rss>, viewed 30 May 2011.

81 Pers. comm., 11 March 2011.

82 Emu, B., et al., 'HLA Class I-Restricted T-Cell Responses May Contribute to the Control of Human Immunodeficiency Virus Infection, but Such Responses Are Not Always Necessary for Long-Term Virus Control', 2008, *Journal of Virology*, vol. 82, p. 5398.

83 CCR5, Trim5 alpha, Apobec3G, KIR.

84 This contingent includes defensive cells that patrol our tissues and bloodstream like macrophages, natural killer cells and dendritic cells. Vaccine-makers have ignored them because, unlike antibody-producing B-cells and T-cells, they cannot be trained. The innate immune system may not be smart, but it appears to play a crucial role in alerting and coordinating the various arms of the immune system for a rapid response—and a rapid response may be of the essence in controlling HIV. One thing all controllers do is to knock down the levels of virus very quickly. Some gene-mining studies of controllers are showing that the genes of the innate immune system (like KIR) can be highly protective against HIV. Bashirova, A.A., et al., 'HLA/KIR Restraint of HIV: Surviving the Fittest', 2011, *Annual Review of Immunology*, vol. 29, p. 295.

85 Hear a presentation by Frank Plummer, Chief Science Advisor and Scientific Director of the Public Health Agency of Canada's National Microbiology Laboratory in Winnipeg, Manitoba, who has studied the Kenyan prostitutes for 17 years, uploaded on TED, 11 March 2011, <youtube.com/watch?v=zGiJZwBXJ6Y>, viewed 30 May 2011.

86 In 2009, The University of Botswana opened a medical school, in collaboration with the University of Melbourne, Australia; see <uninews.unimelb.edu.au/news/1252/>, viewed 30 May 2011.

87 I had met Daniel in Melbourne four years earlier. As a Monash University medical student, he had been inspired by emeritus professor Roger Short (a vet who specialises in human and animal reproduction) to go forth and use his talents to help the world. Short's evangelism fell on willing ears. Daniel commenced specialist physician training in Australia, worked on AIDS projects in Thailand, and from 2006 to 2008 was an internal medicine specialist at Princess Marina Hospital in Gaborone, Botswana, a job the University of Pennsylvania employed him for. He will complete training in infectious diseases in Australia in 2012. His wife, Kristi Roberts, is also one of Roger's missionaries. She first went to Botswana to study the feeding habits of warthogs, endearing creatures that they are. Then, enamoured with Botswana, she returned to Roger for another mission. He suggested she try and tackle the bewildering phenomenon that despite the incessant public campaigns and the free availability of condoms, HIV was stubbornly continuing to spread among the youth of Botswana, particularly young women. Kristi decided to borrow a technique that had been used to good effect in Cambodia to raise consciousness about the plight of women's daily kilometre-long marches to get water from wells. The technique involves equipping people with cameras, and encouraging them to document their own lives. The subsequent photo exhibition serves as a mirror forcing an examination of their choices; and a lens focusing the wider community on the issue. During one of Kristi's visits to Roger's tiny office she met Daniel. Roger told me it was as if he witnessed Cupid's arrow: love at first sight. The rest is a 'happy ever after' story: Daniel and Kristi pursued their missions in Gaborone, and are now the doting parents of two baby girls, Allegra Aurora Loapi and Juliette Savuti. My visit to Gaborone was primarily to interview the researchers at Harvard Botswana. But no-one there could see me until the Monday. It was Friday in Gaborone, and Daniel was eager for me to shadow him in the wards.

88 To protect his privacy I have not used his real name. Daniel Stefanski suggested I use the name Tau. It means brave or lion in Setswana.

89 Setswana is the language spoken in Botswana.

90 Interview with Max Essex, April 2008.

91 UN population estimates in 2002 were that life expectancy would plummet from 65 years in 1990–1995 to 39.7 years in 2000–2005. By 2007, the UN estimate had increased to 50, probably reflecting HIV drug treatment; see <avert.org/aids botswana.htm>, and see also see the 2007 wall chart from the UN on AIDS statistics and life expectancy, <un.org/esa/population/publications/AIDS_Wallchart_web_2007/Population%20and%20HIV-AIDS%202007.htm>, viewed 30 May 2011.

92 Farley, M., 'At AIDS Disaster's Epicenter, Botswana is a Model of Action; During U.N. conference, Leader Speaks of National "Extinction," but Country Plans Continent's Most Ambitious Programs', 2001, Los Angeles Times, 27 June 2001; <aegis.com/news/lt/2001/LT010621.html>, viewed 8 July 2011.

93 UN AIDS Report on the Global AIDS Epidemic, 2010, p. 181; previously available at <unaids.org/globalreport/Global_report.htm>, not accessible 8 July 2011.

94 UN Aids 2006 Epidemic Update; previously available at <data.unaids.org/pub/ .../2007/crp_sgtf_on_women_girls_hiv_aids_en.pdf>, not accessible 8 July 2011.

95 Cohen, J., 'Botswana's Success Comes at Steep Cost', 2008, *Science*, vol. 321, p. 526.

96 The WHO endorses recent studies that show circumcision can reduce heterosexual HIV transmission by 60%; see <who.int/hiv/mediacentre/news68/en/index.html>, viewed 30 May 2011.

97 The figures on HIV prevalence among adults in these countries are from 2009; see p. 188 of <unaids.org/globalreport/Global_report.htm>, viewed 30 May 2011. Interview with Max Essex, Boston, April 2008.

98 When the first cases of AIDS emerged in 1981, Essex, like Steve O'Brien and Robert Gallo, was studying cancer-causing viruses. Essex focused on a cancer-causing virus of cats, feline leukaemia virus. This work on the cat virus helped crack the mystery of AIDS. Gallo and Montagnier had each managed to isolate a virus from AIDS patients, but the world needed to be convinced that a virus could indeed cause a collapse of the immune system—the key symptom of AIDS. In 1983, Essex showed that in cats infected with the feline leukaemia virus, their first symptoms were indeed a collapse of the immune system. In fact, as if to pave the way, Essex's paper on immune suppression by the cat virus appeared in *Science* in 1983 just ahead of the papers by Gallo and Montagnier announcing their discovery of a virus suspected of causing AIDS. Peter Duesberg, a professor of molecular and cell biology at the University of California at Berkeley, continues to doubt that AIDS is a result of HIV infection.

99 Kanki, L., et al., 'New Human T-lymphotropic Retrovirus Related to Simian T-lymphotropic Virus Type-111', 1986, *Science*, vol. 232, p. 238.

100 It turns out that there are multiple subtypes of HIV (a difference of greater than 30% in the genetic code designates a new subtype), and to some extent the subtype of HIV corresponds to the character of the epidemic.

101 Soto-Ramirez, L.E., et al., 'HIV-1 Langerhans' Cell Tropism Associated With Heterosexual Transmission of HIV', 1996, *Science*, vol. 271, p. 1291. A recent paper published by Essex and colleagues also sheds more light on why HIV-subtype C is more infectious than HIV-subtype B. After infection with HIV-1B, the levels of virus in the blood and sperm fall within weeks. But after infection with HIV-1C, the virus levels in the blood and sperm stay high for months. <aids.harvard.edu/news/spotlight/v8i2_surprise_finding.html>, viewed 7 August 2011.

102 Royce, R.A., 'Sexual Transmission of HIV', 1997, *New England Journal of Medicine*, vol. 336, p. 1072; Gray, R.H., et al., 'Probability of HIV-1 Transmission per Coital Act in Monogamous, Heterosexual, HIV-1-discordant Couples in Rakai, Uganda', 2001, *The Lancet*, vol. 357, p. 1149.

103 Wikipedia gives an overview and an extensive list of references: <en.wikipedia.org/wiki/Origin_of_AIDS>, viewed 30 May 2011.

104 Gibbons, A., 'Africans' Deep Genetic Roots Reveal Their Evolutionary Story', 2009, *Science*, vol. 324, p. 575; and <sciencedaily.com/releases/2009/04/090430144524.htm>, viewed 30 May 2011.

105 <sciencedaily.com/releases/2009/04/090430144524.htm>, viewed 30 May 2011.

106 <nature.com/news/2008/080716/full/news.2008.948.html>, viewed 30 May 2011.

107 The mitochondrial enzyme that is susceptible is called DNA polymerase gamma. This side effect is seen with a particular class of NRTIs called D-drugs (didanosine, stavudine, zalcitabine); see <atdn.org/simple/mito.html>, viewed 30 May 2011.

108 Wester, C.W., et al., 'Higher-than-expected Rates of Lactic Acidosis Among Highly Active Antiretroviral Therapy-treated Women in Botswana: Preliminary Results From a Large Randomized Clinical Trial', 2007, *Journal of Acquired Immune Deficiency Syndrome*, vol. 46, p. 318.

109 American and European studies on genetic susceptibility to HIV of course included people of African descent, but these were largely people from West Africa and they had 'mixed' their genes with their local population. Interview with Max Essex, April 2008.

110 Globally 36% of adults in need of AIDS treatment receive it; for children the figure is 28%; see <unaids.org/en/resources/presscentre/pressreleaseandstatement archive/2011/june/20110603praids30/>, viewed 8 July 2011.

111 Max Essex estimated that it will take 12 to 15 years to produce a new type of vaccine to trigger the production of neutralising antibodies. It relies on sophisticated chemistry to produce a synthetic fragment of the virus that freezes a vulnerable domain that is normally only transiently exposed. Once antibodies learn to lock onto this domain, they should be able to neutralise the virus (pers. comm., 30 June 2008). Bruce Walker commented, 'Even if we knew exactly what to put in a vial, it would still take ten years'. Pers. comm., 4 May 2010.

One bright spot on the horizon is that drugs may not only be useful in treating infection, they may also be able to prevent it. For instance according to a trial in 2010, the use of the drug tenofovir (which halts HIV replication) in vaginal gels reduced HIV infection rates in women by 40%. Karim, Q.A., et al., 'Effectiveness and Safety of Tenofovir Gel, an Antiretroviral Microbicide, for the Prevention of HIV Infection in Women', 2010, *Science*, vol. 329, p. 1168. And in May 2011, the results of the HPTN 052 trial showed that people taking HIV drugs effectively stop transmitting the virus. 'Antiretroviral therapy is a bigger game-changer than ever before—it not only stops people from dying, but also prevents transmission of HIV to women, men and children,' said Michel Sidibé, UNAIDS Executive Director; see 'Aids at 30: Nations at the Crossroads', <unaids.org/en/resources/presscentre/ pressreleaseandstatementarchive/2011/june/20110603praids30/>, viewed 8 July 2011.

112 For the latest statistics on HIV, see ibid.

Chapter 6—Feeding Nine Billion

1 'The narrowly defined Asian Green Revolution [lasted from] 1965–1990 ... and the resultant increases in food production pulled the region back from the edge of an abyss of famine and led to regional food surpluses within 25 years. It lifted many people out of poverty, made important contributions to economic growth, and saved large areas of forests, wetlands and other fragile lands from conversion to cropping.' Hazell, P.B.R., 'The Asian Green Revolution', IFPRI Discussion Paper, November 2009, IFPRI, Washington, DC. Also see Hesser, L., *The Man Who Fed the World*, 2006, Durban House Publishing, Dallas: 'Within seven years, the

national average yields had doubled', p. 57 (referring to Mexico, the birthplace of the Green Revolution); 'Pakistan doubled wheat production and achieved self-sufficiency … by 1968', p. x; '[India reached] self-sufficiency in wheat in 1972 and all cereals by 1974', p. x.

2 In 1798, Thomas Malthus declared, 'The power of population is indefinitely greater than the power in the earth to produce subsistence for man'. Malthus, T.R., *An Essay on the Principle of Population*, 1798, Oxford World's Classics, Oxford, p. 13; see <en.wikipedia.org/wiki/Malthusian_catastrophe>, viewed 13 July 2011.

   Paul Ehrlich predicted the same thing 170 years later; see Ehrlich, P.R., *The Population Bomb*, 1968, Buccaneer Books, Catchogue, NY. Early editions of the book began with the statement: 'The battle to feed all of humanity is over. In the 1970s hundreds of millions of people will starve to death in spite of any crash programs embarked upon now. At this late date nothing can prevent a substantial increase in the world death rate …'. See <en.wikipedia.org/wiki/The_Population_ Bomb#cite_note-Population_Bomb-3>, viewed 13 July 2011.

3 'This year, according to the Food and Agriculture Organization of the United Nations (FAO), there are sufficient surplus stocks of wheat, rice and coarse grains to feed every person in the world for two months—and that's not counting the food that will actually be eaten in 1985.'

   'Feast and Famine', September 1985, *New Internationalist*; <newint.org/ features/1985/09/05/feast/>, viewed 13 July 2011.

4 After the food crisis of 2008, prices fell and then spiked again in 2011 by 25%. In Egypt, the price of wheat rose by 50% just before the toppling of Mubarak in February 2011; see Parker, J., 'The 9 Billion-people Question: A Special Report on Feeding the World', 24 February 2011, *The Economist*, <economist.com/ node/18200618>, viewed 13 July 2011.

5 Approximately 1.1 billion people have an income of less than a dollar a day. Almost two-thirds live in rice-growing countries of Asia. They spend 30–40% of their income on rice alone; see <beta.irri.org/solutions/index.php?option=com_conten t&task=view&id=366&Itemid=257>, viewed 13 July 2011.

6 Finkel, E., 'Black Harvest: The Battle Against Wheat Rust', June 2009, *Cosmos*, 27, <cosmosmagazine.com/features/print/2819/black-harvest>, viewed 13 July 2011.

7 About 50% of the calories consumed by the world population originate from three cereals: rice (23%), wheat (17%) and maize (10%); see Khush, G.S., 'Productivity Improvements in Rice', 2003, *Nutrition Reviews*, vol. 61, p. 114.

8 The yield growth rate of wheat between 1990 and 2007 was 0.52% per year. The same figure for rice was 0.96%. In the period between 1961 and 1990, the comparable figure for wheat was 2.95%, and 2.19% for rice; see Alston, J.M., et al., 'Agricultural Research, Productivity, and Food Prices in the Long Run', 2009, *Science*, vol. 325, p. 1209, Table S1. Furthermore, 'Between 1987 and 1997 China increased its average rice yields from 5.4 t/ha to 6.4 t/ha, yet between 1997 and 2007 no further clear increase has been achieved.' Zhu, X.G., et al., 'Improving Photosynthetic Efficiency for Greater Yield', 2010, *Annual Review of Plant Biology*, vol. 61, p. 235.

9 Ibid.

10 For a description of Borlaug's doggedness in introducing new wheat varieties to India, see Hesser, L., *The Man Who Fed the World*, 2006, Durban House Publishing, Dallas, p. 76.

11 <fao.org/news/story/en/item/35686/icode/>, viewed 13 July 2011.

12 'Wheat plants seem to have gone as far as they can shrinking their stalks and leaves to divert more than half their resources to grain' refers to the maximum theoretical Harvest Index—the maximum percentage of its mass that the plant can locate in its grain. It is estimated to be around 50% by Sayre, K.D., et al., 'Yield Potential in Short Bread Wheats in Northwest Mexico', 1997, *Crop Science*, vol. 37, p. 36. The assertion that wheat seems to be close to its maximum Harvest Index comes from Robert Furbank, pers. comm., 21 March 2011, and is reiterated by Parry, M.J., et al., 'Raising Yield Potential of Wheat. II. Increasing Photosynthetic Capacity and Efficiency', 2011, *Journal of Experimental Botany*, vol. 62, p. 453.

13 Sheehy, J.E., et al., 'How the Rice Crop Works and Why it Needs a New Engine', in J.E. Sheehy, P.L. Mitchell and B. Hardy (eds), *Charting New Pathways to C4 Rice*, 2008, World Scientific Publishing, Singapore and International Rice Research Institute, Los Baños, Phillipines, p. 19.

14 Sage, R.F., et al., 'The C4 Plant Lineages of Planet Earth', 2011, *Journal of Experimental Botany*, vol. 62, p. 3155.

15 Photosynthesis is estimated to have begun around three billion years ago. Buick, R., 'When Did Oxygenic Photosynthesis Evolve?', 2008, *Philosophical Transactions of the Royal Society London Biological Sciences*, vol. 363, p. 2731, <ncbi.nlm.nih.gov/pmc/articles/PMC2606769/?tool=pmcentrez>, viewed 22 June 2011.

16 Rye, R., et al., 'Atmospheric Carbon Dioxide Concentrations Before 2.2 Billion Years Ago', 1995, *Nature*, vol. 378, p. 603.

17 Pagani, M., et al., 'Marked Decline in Atmospheric Carbon Dioxide Concentrations During the Paleogene', 2005, *Science*, vol. 309, p. 600. Carbon dioxide levels probably plummeted because the Himalayas rose up, exposing silica-bearing rocks. They sucked carbon dioxide out of the air to form silicon dioxide and carbonates that deposited in the sea as limestone. As explained to me by palaeobotanist Rowan Sage at the University of Toronto, pers. comm., 30 March 2011.

18 Vicentini, A., 'The Age of the Grasses and Clusters of Origins of C4 Photosynthesis', 2008, *Global Change Biology*, vol. 14, p. 2963.

19 In most crop plants, the four-carbon chain is malate.

20 An enzyme called PEP carboxylase is used to knit carbon dioxide into malate. An enzyme called C4 acid decarboxylase releases carbon dioxide from malate. These enzymes are present in the C3 plants but have a low activity. C4 plants raise the activities of these and at least ten other intrinsic genes to very high levels, some more than a 100 times greater than in C3 plants. As explained to me by Robert Furbank, pers. comm., 7 April 2011.

21 See Byrt, C.S., et al., 'C4 Plants as Biofuel Feedstocks: Optimising Biomass Production and Feedstock Quality From a Lignocellulosic Perspective', 2010, *Journal of Integrative Plant Biology*, vol. 53, p. 120; and Somerville, C., et al., 'Feedstocks for Lignocellulosic Biofuels', 2010, *Science*, vol. 329, p. 790. Elephant grass is quoted here as yielding 88 tons to the hectare.

Robert Furbank estimates that wheat performing at its best (for instance, in parts of Argentina and Peru and in a good year at Rothamsted experimental station, UK) would produce 15 tons to the hectare of grain, which corresponds to 30 tons to the hectare of dry biomass. Pers. comm., 12 July 2011.

22 Reynolds, M., et al., 'Raising Yield Potential of Wheat. I. Overview of a Consortium Approach and Breeding Strategies', 2011, *Journal of Experimental Botany*, vol. 62, p. 439.

23 Sheehy, J.E., et al., 'How the Rice Crop Works and Why it Needs a New Engine', in J.E. Sheehy, P.L. Mitchell and B. Hardy (eds), *Charting New Pathways to C4 Rice*, 2008, World Scientific Publishing, Singapore and International Rice Research Institute, Los Baños, Phillipines, p. 19.

24 They included Hei Leung and Richard Brusciewisch from the International Rice Research Institute in the Phillipines, Jane Langdale from Oxford University, Julian Hibberd from Cambridge University; Richard Leegood from Sheffield University, Peter Westoff from Dusseldorf University, Rowan Sage from Toronto University, Gerry Edwards from Washington State University, Tom Brutnell from Cornell University, and Susanne von Caemmerer from Australian National University.

25 His given name is Marshall Hatch.

26 These 'solar panels' are light-harvesting molecules called P680 and P700.

27 'The share of agriculture in official development assistance (ODA) declined sharply over the past two decades, from a high of about 18 percent in 1979 to 3.5 percent in 2004. It also declined in absolute terms, from a high of about $8 billion (2004 US$) in 1984 to $3.4 billion in 2004. The bigger decline was from the multilateral financial institutions, especially the World Bank. In the late 1970s and early 1980s the bulk of agricultural ODA went to Asia, especially India, in support of the green revolution, although this declined dramatically thereafter. Total ODA to agriculture in Africa increased somewhat in the 1980s, but it is now back to its 1975 level of about $1.2 billion.'

    'Agriculture For Development', World Bank Development Report, 2008, World Bank, Washington, DC, p. 41.

28 'Grains go on an even bigger "health kick"', CSIRO Media Release, 8 May 2007, <csiro.au/news/GrainBasedFoods.html>, viewed 22 June 2011.

29 Finkel, E., 'Richard Richards Profile: Making Every Drop Count in the Buildup to a Blue Revolution', 2009, *Science*, vol. 323, p. 1004.

30 The two genes that dwarfed the Green Revolution wheats were called 'Rht1' and 'Rht2'. The names stand for reduced height trait 1 and 2; see <en.wikipedia.org/wiki/Norin_10_wheat>, viewed 22 June 2011.

31 Ashikari, M., et al., 'Cytokinin Oxidase Regulates Rice Grain Production', 2005, *Science*, vol. 309, p. 741.

32 Gene markers are variable parts of genomes that can be used to track different genes. They are usually repetitive sequences known as Restriction Fragment Length Polymorphisms or RFLPs.

33 LIDAR, which stands for Light Detection And Ranging, is 'remote sensing' technology that detects objects by bombarding them with pulses from laser beams. It's the same technology that's used in a police speed gun.

34 The temperature of a plant can reveal many things:

    It is a rough guide to the rate of photosynthesis, because a cool plant must be losing water vapour from open stomata (mouths). If the stomata are open, it means the plant is also breathing in carbon dioxide, which means it is photosynthesising. So, cooler plants may be doing more photosynthesis than warmer plants.

It is a guide to salt tolerance, because salt-intolerant plants tend to stop drawing water, making them hotter.

It is a guide to water-efficient plants, because plants that are good at drawing water from their roots stay cooler.

So how do the plant researchers decide which of these features is being tracked by the plant's temperature? They make the other two irrelevant. So, if plants are growing in plenty of water and low salt, then differences in temperature reflect their rate of photosynthesis.

35 Fukayama, H., et al., 'Activity Regulation and Physiological Impacts of Maize C_4-specific Phosphoenolpyruvate Carboxylase Overproduced in Transgenic Rice Plants', 2003, *Photosynthesis Research,* vol. 77, p. 227.

36 International C_4 Rice Consortium, International Rice Research Institute site: <irri.org/c4rice>, viewed 22 June 2011; CSIRO site: <csiro.au/science/C4-rice-consortium.html>, viewed 22 June 2011.

37 All these reactions take place in the chloroplasts that reside inside plant cells. These are tiny oval structures that probably originated as symbiotic purple-green bacteria.

38 Pers. comm., 30 March 2011.

39 Kajala, K., et al., 'Strategies for Engineering a Two-celled C4 Photosynthetic Pathway into Rice', 2011, *Journal of Experimental Botany,* vol. 62, p. 3001; and Sheehy, J.E., Mitchell, P.L. and Hardy, B. (eds), *Charting New Pathways to C4 Rice,* free book, <books.google.com.au/books?id=dHESDZQQ8AUC&printsec=frontcover&dq=c leome+C4+agriculture&source=bl&ots=E6EkcImqwG&sig=ohf3slu8HFtdjZNiG ES8hfIQ9ao&hl=en&ei=tRORTd-1OoySuwOP5MGaAQ&sa=X&oi=book_result &ct=result&resnum=3&ved=0CCYQ6AEwAg#v=onepage&q=cleome%20C4%20 agriculture&f=false>, viewed 22 June 2011.

40 Brautigam, A., 'An mRNA Blueprint for C4 Photosynthesis Derived From Comparative Transciptomics of Closely Related C3 and C4 Species', 2011, *Plant Physiology,* vol. 55, p. 142.

41 Julian Hibberd at Cambridge University has inserted an RNAi gene into rice that deactivates Rubisco only in mesophyll cells. As conveyed to me by Robert Furbank, pers. comm., 21 March 2011.

42 C_4 plants use CO_2 so efficiently that they keep their stomata—their breathing pores—closed more often than their C_3 relatives. Like a dog that doesn't pant, this also makes them warmer than their C_3 relatives.

43 Pers. comm., 21 March 2011.

44 Reynolds, M., et al., 'Raising Yield Potential of Wheat. I. Overview of a Consortium Approach and Breeding Strategies', 2011, *Journal of Experimental Botany,* vol. 62, p. 439.

45 The truly remarkable nature of wheat is not visible to the naked eye. As mammals, we are accustomed to the idea that chromosomes comes in duplicate, that is, we have two copies of each of our 23 chromosomes. Wheat, however, has six copies of each of its seven chromosomes! How did such a thing happen? Botanists believe that wheat is the result of a mating between three sets of parents; three different grasses contributed their genomes to make wheat. Two are still growing wild in the Middle East: goat grass and Einkorn wheat.

It seems to have happened in two stages. About 10,000 years ago, somewhere in south-east Turkey, the ancestor of Einkorn wheat (genome AA) and another grass (genome BB) combined their chromosomes to create Emmer wheat (AABB), a wheat with four copies of each chromosome. Modern-day durum or pasta wheat also has four copies of each chromosome. Then, about 9,000 years ago in the Near East, another species of goat grass, *Aegilops tauschii*, got into the act, contributing its genome (DD) to form bread wheat (AABBDD). See Shewry, P.R., 'Wheat', 2009, *Journal of Experimental Botany*, vol. 60, p. 1537, <jxb.oxfordjournals.org/content/60/6/1537.full>, viewed 13 July 2011.

46 Robert Furbank, pers. comm., 21 March 2011. In a longer-term approach, Martin Parry and colleagues at the Rothamsted Centre for Crop Genetic Improvement in Hertfordshire, UK, and colleagues at the Australian National University in Canberra are trying to turbo-charge Rubisco. But not by copying C_4 plants. They are trying a trick used by purple-green bacteria. These bacteria import bicarbonate molecules into their cells. Once inside the soupy interior, CO_2 bubbles out of the dissolved bicarbonate like gas from a fizzy drink. The genetic engineers are trying to transfer these bacterial bicarbonate importers into wheat; see Parry, M.J., et al., 'Raising Yield Potential of Wheat. II. Increasing Photosynthetic Capacity and Efficiency', 2011, *Journal of Experimental Botany*, vol. 62, p. 453.

47 Film about Borlaug on YouTube; see <youtube.com/watch?v=m2TmEdiXTvc&feature=related>, viewed 22 June 2011.

Chapter 7—Meet Your Ancestor

1 Darwin, C., *On the Origin of Species by Means of Natural Selection or the Preservation of Favoured Races in the Struggle for Life*, 1859, John Murray, London, Random House ed., 1993, p. 649. The entire quote reads: 'There is grandeur in this view of life, with its several powers, having been originally breathed by the Creator into a few forms or into one; and that, whilst this planet has gone cycling on according to the fixed law of gravity, from so simple a beginning endless forms most beautiful and most wonderful have been, and are being evolved.'

2 The prefix Ur is an allusion to humans' first attempt at civilisation: the ancient Mesopotamian city of Ur. Likewise, the Ur-metazoan was the first attempt at sophisticated multicellular life.

3 Syed, T. and Schierwater, B., '*Trichoplax adhaerens*: Discovered as a Missing Link, Forgotten as a Hydrozoan, Re-discovered as a Key to Metazoan Evolution', 2002, *Vie Milieu*, vol. 52(4), p. 177; <ecolevol.de/pubs/2002/syed-schierwater-VM2002b.pdf>, viewed 22 June 2011.

4 As related to me by Vicki Pearse, a *Trichoplax* expert at the University of California, Institute of Marine Sciences, Santa Cruz, California, pers. comm., 14 December 2010.

5 Srivastava, M., et al., 'The Trichoplax Genome and the Nature of Placozoans', 2008, *Nature*, vol. 454, p. 955.

6 Most of the time *Trichoplax* reproduces by splitting or budding, that is, not sexually. But it may have sex. When colonies get crowded, researchers have seen it produce something that looks like an egg, although it always seems to die before hatching into anything and no-one has found a sperm; pers. comm., Vicki Pearse, 14 December 2010.

7 Mikhailov, A.T. and Gilbert, S.F., 'From Development To Evolution: The Re-establishment of the "Alexander Kowalevsky Medal"', 2002, *International Journal*

of Developmental Biology, vol. 46, p. 693; <ijdb.ehu.es/web/pdfdownload.php? doi=12216980>, viewed 16 July 2011.

8 Two of the people I interviewed had different views about the evolutionary origins of *Trichoplax. Trichoplax* expert Vicki Pearse told me, 'A sea squirt remains quite a complex animal even after its metamorphosis from a swimming tadpole stage; it has a full gut, beating heart, gonads. A placozoan [like *Trichoplax*] has no organs at all, so both genetic and morphological evidence keep it close to the base of the animal tree … I cannot think of any evidence (or any argument) that supports its having become simplified from what I'd call an advanced animal.' Pers. comm., 14 December 2010. Eldon Ball, an evolutionary biologist from Australian National University disagrees: 'It seems to me that [most people] have concluded that *Trichoplax* is secondarily simple', pers. comm., 14 January 2011, by which he means that he believes it is most likely to have descended from a more advanced ancestor.

9 That's not to say that sponges don't display huge diversity: 'everything from nondescript lumps on rocks to the beautiful complex structure of Venus's Flower Basket', as Eldon Ball points out.

10 Fahey, B. and Degnan, B.M., 'Origin of Animal Epithelia: Insights From the Sponge Genome', 2010, *Evolution & Development,* vol. 12, p. 601. Bernie Degnan, a sponge biologist at the University of Queensland, points out that not all cells of the sponge are simple. Some can link up to form an 'epithelial-like layer'. And the larvae carry a pigment ring that responds to light. Pers. comm., 13 April 2011.

11 Maclennan, A.P. and Dodd, R.Y., 'Promoting Activity of Extracellular Materials on Sponge Cell Reaggregation', 1967, *Journal of Embryology and Experimental Morphology,* vol. 17, p. 473.

12 Becoming multicellular appears to be a survival strategy in the face of predation. The single-celled algae *Chlorella vulgaris* prefers the single life. But with the introduction of a predator, after a hundred generations the algae went multicellular. Boraas, M.E., 'Phagotrophy by a Flagellate Selects for Colonial Prey: A Possible Origin of Multicellularity', 1998, *Evolutionary Ecology,* vol. 12, p. 153.

13 <choano.org/wiki/Choanoflagellates>, viewed 22 June 2010.

14 Mann, A., 'Sponge Genome Goes Deep', 4 August 2010, *Nature,* <nature.com/ news/2010/100804/full/466673a.html>, viewed 22 June 2010.

15 Genomes have not really clarified which species is closest to the ancestral state. Comb jellies (Ctenophores) have a nerve net, so they have a more complex body than does the sponge. Yet their genome is simpler than that of the sponge. If the genome is the true guide to complexity, it may be that the sponge is not an early model animal but a late model that has been simplified. So, taking its features as 'ancestral' could give us a skewed notion of the path of evolution—just as if you took a current 'retro' version of the Mini Minor as an example of an old-style car, you'd be sorely mistaken if you assumed airbags were part of the package of 1960s cars. The bottom line is you can have simple animals with complex genomes (sponges and *Trichoplax*) or you can have complex animals with simple genomes (Ctenophores). Scientists are still battling over whether genomes or bodies are the best guide as to where an animal should sit on the family tree. As Vicki Pearse put it, 'It is not the overall complexity of the genome—or the body—that indicates an earlier lineage or a later branching one. It is accurately tracing the changes (losses, duplications or substitutions) in specific bits of the genome that will tell us the

story—and we are not yet quite able to do that in a consistent way. Different genes seem to give different results, different analyses from different labs seem to give different results, etcetera. So complex, frustrating, tantalizing! But some of it seems to be taking shape …'. Pers. comm., 16 January 2011. Also, see Miller, G., 'On the Origin of the Nervous System', 2009, *Science*, vol. 325, p. 24.

16 Pers. comm., 14 December 2010.

17 Pers. comm., 19 October 2010.

18 This bacterium was mistakenly believed to be the cause of influenza, which we now know to be caused by a virus. However, it held onto its name. *Haemophilus influenzae* type b (Hib) is responsible for severe pneumonia, meningitis and other diseases almost exclusively in children under five years of age. A vaccine is now part of the infant immunisation schedule in over one hundred countries. See <who.int/immunization/topics/hib/en/index.html>, viewed 22 June 2011.

19 The latest estimate is 22,500 from the curator of OMIM, Ada Hamosh, pers. comm., 3 May 2011.

20 Miller, D.J. and Ball, E.E., 'The Gene Complement of the Ancestral Bilaterian: Was Urbilateria a Monster?', 2009, *Journal of Biology*, vol. 8, p. 89; <jbiol.com/content/8/10/89>, viewed 22 June 2011.

21 David Miller developed his views through close collaboration with evolutionary biologist Eldon Ball at Australian National University, Melbourne University geneticist Rob Saint and geneticist Walter Gehring at the University of Basel in Switzerland.

22 See Carroll, S.B., *Endless Forms Most Beautiful: The New Science of Evo Devo and the Making of the Animal Kingdom*, 2005, Phoenix, London, p. 71.

23 Pashmforoush, M., et al., 'Nkx2-5 Pathways and Congenital Heart Disease: Loss of Ventricular Myocyte Lineage Specification Leads to Progressive Cardiomyopathy and Complete Heart Block', 2004, *Cell*, vol. 117, p. 373; and see Jeffrey, S., '"Tinman" Gene Focus of Wall Street Journal Feature', 2004, *Heartwire* (available online).

24 Travis, J., 'Eye-opening Gene', 10 May 1997, *Science News*, <sciencenews.org/sn_arc97/5_10_97/bob1.htm>, viewed 22 June 2010.

25 Slack, J., 'A Rosetta Stone for Pattern Formation in Animals?', 1984, *Nature*, vol. 310, p. 364.

26 Archaeologists were clueless to the meaning of Egyptian hieroglyphics until 1799 when, during Napoleon's campaign in Egypt, a soldier discovered a stone slab in the town of Rosetta, Egypt, that was inscribed in Egyptian hieroglyphics, Egyptian demotic and ancient Greek. It was a decree by Ptolemy V in 196 BC, setting out, among other things, a tax exemption for the priests and also that the decree should be hammered out in the three languages. The ancient Greek allowed archaeologists to infer the meaning of the hieroglyphic text.

27 Slack, J., 'A Rosetta Stone for Pattern Formation in Animals?', 1984, *Nature*, vol. 310, p. 364.

28 Pers. comm., 1 September 2010.

29 Kortschak, R., et al., 'EST Analysis of the Cnidarian *Acropora millepora* Reveals Extensive Gene Loss and Rapid Sequence Divergence in the Model Invertebrates', 2003, *Current Biology*, vol. 13, p. 2190. One set of 'toolkit' genes that is well represented in humans and corals but not flies and worms, is the wnt gene set. They mould developing tissues and when faulty, that can lead to wingless flies,

or limbless humans (Niemann, S., et al., 'Homozygous WNT3 Mutation Causes Tetra-amelia in a Large Consanguineous Family', 2004, *American Journal Human Genetics*, vol. 74, p. 558). So researchers were not surprised to discover that the human genome contained 12 members of the wnt gene set, while flies had seven and worms had five. It seemed to correlate with the complexity of the animals' tissues. But they got a shock when they found coral had 11 members of the wnt set—and that they were more similar to human wnts than to the fly or worm wnt genes. The story holds true for about 10% of the coral genes: they have clear matches to human genes but not to those in fruit flies or roundworms. Other examples of genes that are present in coral and humans but not worms and flies include: Dikkopf (an antagonist of the wnt pathway), Noggin and Gremlin (antagonists of the TGF beta pathway involved in axis specification and tissue moulding); Churchill and Tumorhead (they direct nervous system development), Retinol dehydrogenase 8 (involved in retinol recycling), DNMT3 and MeCpG binding proteins (involved in DNA methylation). Ibid and David Miller, pers. comm., 18 July 2011.

30 Pers. comm., 18 August 2010.

31 The so-called 'universal stress proteins' (USP) were known primarily from bacteria. In more recent times researchers had found them in fungi and plants but until Miller found them in coral, never in any animal. Another couple of exotic genes were thought to be exclusive to plants before they showed up in the coral genome. These were the oxylipin synthesis gene, which produces jasmonates, responsible for the fruity odours of plants, and a gene that codes for the synthesis of abscisic acid, the quintessential plant SOS signal. See Foret, S., et al., 'Phylogenomics Reveals an Anomalous Distribution of USP Genes in Metazoans', 2011, *Molecular Biology and Evolution,* vol. 28, p. 153.

32 Miller told me that they collect coral on the afternoon of the predicted spawning. In their laboratory tank, egg and sperm bundles float to the surface, break up and fertilise each other. Within this tangle of egg and sperm, usually only those of the same species will fertilise each other. Like plants, each coral animal releases both eggs and sperm, but these cannot fertilise each other; they are said to be self-incompatible. As explained to me by Miller during a visit to his laboratory at James Cook University, Townsville, 18 August 2010.

33 Foret, S., et al., 'Phylogenomics Reveals an Anomalous Distribution of USP Genes in Metazoans', 2011, *Molecular Biology and Evolution,* vol. 28, p. 153.

34 DiNardo, S., Sher, E., et al., 'Two-tiered Regulation of Spatially Patterned *engrailed* Drosophila', 1988, *Nature*, vol. 332, p. 604. (E. Sher is now E. Finkel.)

35 In the fruit fly, the *engrailed* gene designs part of each segment, wing and the nervous system. In human beings it has been co-opted to design the fine features of the nervous system.

36 Darwin, C., *On the Origin of Species by Means of Natural Selection or the Preservation of Favoured Races in the Struggle for Life,* 1859, John Murray, London.

37 Darwin failed to recognise the significance of Mendel's discovery of the gene in 1860. In 1900, three botanists rediscovered the gene.

38 Koonin, E.V., 'Darwinian Evolution in the Light of Genomics', 2009, *Nucleic Acids Research*, vol. 37, p. 1011, <nar.oxfordjournals.org/content/37/4/1011>, viewed 22 June 2011. To read more I suggest Koonin, E.V., *The Logic of Chance: The Nature and Origin of Biological Evolution,* FT Press, New Jersey, 2011, <http://my.safaribooksonline.com/book/-/9780132623117>.

39 Ibid.

40 The Universal Stress Protein genes are absent from chordates (animals with nerve cords), except for one primitive group called *Ciona*. They are not in ecdysozoans (animals that moult, including insects and roundworms or Nematoda) or Placozoans (dear *Trichoplax*). But they are present in lophotrochozoans (flatworms, molluscs, segmented worms; formally known as Platyhelminthes, molluscs, Annelids) and they are in cnidarians (corals, jellyfish, sea anemones) and sponges, plants and fungi. The fruity genes show up in Cnidarians. See Foret, S., et al., 'Phylogenomics Reveals an Anomalous Distribution of USP Genes in Metazoans', 2011, *Molecular Biology and Evolution*, vol. 28, p. 153.

41 Moran, N.A. and Jarvik, T., 'Lateral Transfer of Genes From Fungi Underlies Carotenoid Production in Aphids', 2010, *Science*, vol. 328, p. 624.

42 For different views on the controversy, see Koonin, E.V., 'Darwinian Evolution in the Light of Genomics', 2009, *Nucleic Acids Research*, vol. 37, p. 1011, <nar. oxfordjournals.org/content/37/4/1011>, viewed 22 June 2011; and Kurland, C.G., et al., 'Horizontal Gene Transfer: A Critical View', 2003, *Proceedings of the National Academy of Science USA*, vol. 100, p. 9658, <pnas.org/content/100/17/9658.abstrac t?ijkey=dd4b70539a380404752399cf13f39f7f4be5dd2d&keytype2=tf_ipsecsha>, viewed 22 June 2011.

43 Sponges do not have Hox genes but they do have relatives known as homeobox genes.

44 For a discussion of Hox genes and the Cambrian explosion see Carroll, S.B., *Endless Forms Most Beautiful: The New Science of Evo Devo and the Making of the Animal Kingdom*, 2005, Phoenix, London, Chapter 6, 'The Big Bang of Animal Evolution', p. 137.

45 Richard Fortey, a palaeontologist at the Natural History Museum in London refers to a 10-million-year duration for the Cambrian Explosion, though the feature story examines recent evidence for a much earlier start. Fortey, R., 'The Cambrian Explosion Exploded?', 2001, *Science,* vol. 293, p. 438.

46 Richards, G.S., et al., 'Sponge Genes Provide New Insight into the Evolutionary Origin of the Neurogenic Circuit', 2008, *Current Biology*, vol. 18, p. 1156.

47 Archaea were once classified together with bacteria, but in 1977 Carl Woese discovered they represented a distinct model of life. They are now classified as one of three major domains of cellular life: archaea, bacteria and eukaryotes. See <en. wikipedia.org/wiki/Archaea>, viewed 16 July 2011.

48 Alberts, B., et al., *Molecular Biology of the Cell*, 5th ed., 2008, Garland Science, New York, p. 815.

49 Overbye, D., 'Microbe Finds Arsenic Tasty: Redefines Life', 2 December 2010, <nytimes.com/2010/12/03/science/03arsenic.html>, viewed 22 June 2011; Pennisi, E., 'Discoverer of Arsenic Bacteria, in the Eye of the Storm', 20 December 2010, *ScienceNow*, <news.sciencemag.org/sciencenow/2010/12/arsenic-researcher-asks-for-time.html?ref=hp>, viewed 22 June 2011.

50 Gould, S.J., 'Planet of the Bacteria', 1996, *Washington Post Horizon*, vol. 119(344), p. H1, <stephenjaygould.org/library/gould_bacteria.html>, viewed 22 June 2011.

51 <discovermagazine.com/2004/dec/discover-dialogue>, viewed 16 July 2011; <jcvi. org/cms/press/press-releases/full-text/article/137-marine-microbial-genomes-from-cultured-samples-are-sequenced-and-compared-to-sorcerer-ii-global/>, viewed 16 July 2011. Comment by David Tribe, pers. comm., 15 July 2011.

52 For a description of the imagined sequence of events that led to the evolution of mitochondria and chloroplasts, see Alberts, B., et al., *Molecular Biology of the Cell*, 5th ed., 2008, Garland Science, New York, pp. 26–30, 859–860.

53 Maxmen, A., 'Ancient Macrofossils Unearthed in West Africa', 30 June 2010, *Nature News*, <nature.com/news/2010/100630/full/news.2010.323.html>, viewed 22 June 2011. Monash University palaeontologist Patricia Rich also points out that some multicellular organisms like Horodyskia were around at least 1.4 billion years ago, however 'we are still debating as to whether it was a plant or animal'. Pers. comm., 19 April 2011. See also Fedonkin, M.A., 'The Origin of the Metazoa in the Light of the Proterozoic Fossil Record', 2003, *Paleontological Research*, vol. 7, pp. 9–41.

54 Miller, G., 'On the Origin of the Nervous System', 2009, *Science*, vol. 325, p. 24.

55 'Three Billion-year-old Genomic Fossils Deciphered', *ScienceDaily*, <sciencedaily. com/releases/2010/12/101219140815.htm>, viewed 22 June 2011.

56 Alberts, B., et al., *Molecular Biology of the Cell*, 5th ed., 2008, Garland Science, New York, p. 904.

57 Hazan, R.B., et al., 'Cadherin Switch in Tumor Progression', 2004, *Annals of the New York Academy of Science*, vol. 1014, p. 155; and Desgrosellier, J.S. and Cheresh, D.A., 'Integrins in Cancer: Biological Implications and Therapeutic Opportunities', 2010, *Nature Reviews Cancer*, vol. 10, p. 9.

58 King, N., et al., 'The Genome of the Choanoflagellate *Monosiga brevicollis* and the Origin of Metazoans', 2008, *Nature*, vol. 451, p. 783. It has genes for integrin alpha but not integrin beta. And for another protist genome that carries integrins, see Sebé-Pedrós, A., et al., 'Ancient Origin of the Integrin-mediated Adhesion and Signaling Machinery', 2010, *Proceedings of the National Academy of Science USA*, vol. 107, p. 10142, <pnas.org/content/107/22/10142.full#ref-3>, viewed 16 July 2011. Also for a broader discussion, see the origins of the multicellularity project, <broadinstitute.org/annotation/genome/multicellularity_project/MultiHome. html>, viewed 22 June 2011.

59 Darwin, C., *On the Origin of Species by Means of Natural Selection or the Preservation of Favoured Races in the Struggle for Life*, 1859, John Murray, London, Random House ed. 1999, p. 649.

Index

Note: Page numbers in bold type indicate photos; d after the number indicates a diagram; and p indicates a painting.